Farm Fresh Retailing | Case of Customer Satisfaction

Farm Fresh Retailing

Case of Customer Satisfaction

Authors:

TANVEERA HASSAN LONE

Dr. JYOTI KACHROO

NAVEED HAMID

SATISH SERIAL PUBLISHING HOUSE

403, Express Tower, Commercial Complex
Azadpur, Delhi-110033 (India)
Phone : 011-27672852, Fax : 91-11-27672046
e-mail : info@satishserial.com, hkjain1975@yahoo.com
Website : www.satishserial.com

Published by :

SATISH SERIAL PUBLISHING HOUSE

403, Express Tower, Commercial Complex, Azadpur, Delhi-110033 (INDIA)
Phone : 011-27672852 Fax : 91-11-27672046
E-mail : info@satishserial.com, hkjain1975@yahoo.com

ISBN 978-93-88020-59-6

Composed, Designed & Printed in India

About the Authors

Miss. Tanveera Hassan Lone possesses a meritorious academic record in Masters (Agri-Business Management) programme. She has obtained her degree in MBA (Agri-Business Management) from the Sher-e-Kashmir University of Agricultural Sciences and Technology of Jammu. She has graduated in Forestry from SKUAST K. She has three research publications in referred and NAAS rated Journals, two abstracts and two conferences into her credit. She has worked in International project sponsored by IFPRI. Her field of expertise is Agricultural Business Management, Food Industry Marketing. Presently, she is working with Lovely Professional University, Punjab.

Dr. Jyoti Kachroo has remained gold medalist (Two) and meritorious throughout her academic career, graduated and post-graduate (Economics) from University of Jammu, Jammu and Kashmir. She was initially appointed as lecturer (Economics) in Govt. Degree Collage, Jammu and later joined the division of agricultural Economics and Statistics, SKUAST, Jammu as an Associate Professor. Presently, she is Professor and Head, Division of Agricultural Economics & Agri-Business Management. Her field of expertise is Women empowerment, agricultural production economics, marketing and econometric model based research. She has published about more than 90 articles, which include chapters, popular articles, review papers and about 65 research publications in referred journals. She has been honoured with D.T.Doshi Foundation Award for best presentation. She has more than 26 years experience of teaching and research and she has guided 20 research scholars.

Mr. Naveed Hamid possesses a meritorious academic record and awarded by university merit certificate in recognition of his having achieved the prescribed scholastic standards of high order OGPA in Masters (Agri-Business Management) programme. He has more than 4 years of teaching and research experience in the field of Agricultural Business. He has more than 60 research publications in referred, NAAS rated and UGC approved journals, 22 abstracts, 21 books chapters, several articles, 06 books and so many international/National conferences to his credit. He is recipient of more than 12 International/National recognition award from repute organization for his contribution to the Agri-Business field. Among those awards, he is bestowed with Krishi Bhushan and IRDO Young Fellow award. His field of expertise is Agricultural Business Management, Supply chain/Value chain and marketing. Mr. Naveed Hamid is presently working in Ministry of Agriculture Co-operation and Farmers Welfare. He is also an Ambassador, ACPCL (M.P) for promotion of FPO and Agri-business related activities in Jammu and Kashmir state.

Preface

Being perishable in nature, the annual post-harvest loss in fruits considering quality and quantity is very high ranging from 30-40% of the total production. Retail is one of the largest industries in the world in terms of both size and the number of people employed globally. It is one of the finest industries from the supply chain management perspective. Customer satisfaction is an essential element to get operational efficiency. Customer satisfaction of farm fresh retailing is the basis of company success. Because of the vital role, customer satisfaction plays an indispensible part within the retail organizations in present era.

The various chapters of the book focus on the customer satisfaction in retail industry for fresh farm products and its proper management. The book also throws light on the aspect of retailing scenario, Farm fresh retailing and Wheel of retailing, etc. The book mainly focuses on identification of various factors that leads to the customer satisfaction in retail industry for farm fresh products in the sample area. The various chapters also discuss the retailing scenario in detail along with the wheel of retailing approach. The next chapter throws light on the research methodology, tools adopted for analysis and result/ outcome related to farm fresh retailing in detail.

We felt our heartfelt gratitude to Dr. Dileep Kachroo (Chief Scientist, SKUAST, Jammu), whose genuine guidance and conceptual technical ideas during the course of planning and timely help in finalizing this book.

Contents

Chapter 1

Retailing Background

Retailing Scenario

The word 'retail' means to sell or be sold directly to individuals. Retail is India's largest industry, and arguably the one with the most impact on the population. It is the country's largest source of employment after agriculture, has the deepest penetration to rural India and generates more than 10 per cent of India's GDP (Philip Kotler, 2009). However, retailing in India has so far, been mostly in the hands of small disorganized entrepreneurs. It is also India's least evolved industries. In fact, it is not even considered a real industry (Berman & Evans, 2008). The industry suffers from lack of management talent, poor access to capital, unfavourable regulation and denial of access to best practices. The Indian retail industry is now beginning to evolve in line with the transformation that has swept other large economies. India has some 12 million retail outlets, but many of these act merely as subsistence providers for their owners and survive on a cost structure, where labour and land is assumed to be free and taxes nil. (K. Singh, 2014).

However, only recently has there been an awakening in this sector, with more organized retailers starting to make an impact. The liberalization of the consumer goods industry, initiated in the mid-80s and accelerated through the 90s has begun to impact the structure and conduct of the retail industry. Backed by changing consumer trends, liberalization in mindsets driven by media, new opportunities and increasing wealth, retailing in India, presents a vast opportunity for a variety of businesses , real estate, store design and operations, visual merchandising logistics and communications, B_2C service providers, and FMCG companies, who can add to their offers by partnering this revolution (K. Singh, 2014).

The Indian Retailing Industry stands poised to take off into the 21st century. It is one of the fastest growing sectors in the nation that caters to the world's second largest consumer market. India has five million retailers with a business volume of $180 million growing at 5 to 7 per cent a year (Department of Industrial Policy and Promotion website, Union Budget, 2015–16). The middle class drives retailing anywhere in the world and this segment should have reasonable income. The next driver is availability of variety of goods, products and brands. The third one is "sense of awareness".

In other developing economies, this transformation has already begun. In many of these countries, organized retail already has a 40 per cent share of the market, compared to India's current levels of 2 per cent. As India goes through this transformation, new businesses with sales of 1 billion - 2 billion US $ will be created in grocery and of 250 million – 500 million US $ in apparel. Smaller but still interesting opportunities will be created in other sectors like books, electronics, and music. This transformation will also impact the supply chain in agriculture, the tax collections from trade and the way people shop (McKinsey Global Institute Forecasts, 2016-17).

Retailing is the largest private sector industry in the world economy with the global industry size exceeding $6.6 trillion and a latest survey has projected India as the top destination for retail investors. India is currently the twelfth largest consumer market in the world. The Indian consumer market is likely to grow four times by 2025 (a McKinsey report, the rise of Indian Consumer Market, 2015). A good talent pool, unlimited opportunities, huge markets and availability of quality raw materials at cheaper costs is expected to make India overtake the world's best retail economies by 2042, according to industry players.

The Indian retail industry is the fifth largest in the world. With continued economic expansion and retail growth, India is set to become a US$ 450 billion retail market by 2015, comparable in size to Italy (US$ 462 billion) and much larger than Brazil (US$ 258 billion) today. The present value of the Indian retail market is estimated by the India Retail Report to be around ₹ 12, 00, 000 crores ($ 270 billion) and the annual growth rate is 5.7 per cent. Furthermore, around 15 million retail outlets help India to win the crown of having the highest retail outlet density in the world (V. Aggarwal, 2007).

Retail sector is the largest source of employment after agriculture, and has deep penetration into rural India. It is also believed that 21 million people are employed in the retail sector, which is 7 per cent of the total national workforce, whereas the global average is around

10-12 per cent. It is estimated that an additional eight million jobs will be generated through direct and indirect employment related to the retail sector (A. Sengupta, 2018).

Driven by changing lifestyles, strong income growth and favourable demographic patterns, Indian retail is expanding at a rapid pace. The retail sector of India is going through a fresh period of growth, which is backed by strong economic fundamentals. There has been a massive development of new retail formats, such as malls, hypermarkets, supermarkets and lifestyle stores. The organized sector represents a mere 2 per cent share of this market. It is very low as compared to other developed economies of the world.

Evolutionary Aspect in Retailing

The following table states clearly the way the retail industry has evolved and what changes can be expected in retailing in the future as compared to what it was in the past:

Table 1.1: Comparison of Past and Future

Past	Future
Stability and consolidation	Speed and imagination
TQM six sigma	Design management
Delegation	Abdication
Hierarchy	Seamlessness
Mass production	Personalization
Technology supports change	Technology drives change
Enforce order	Thrive in chaos
Transactions	Relationships

Source: International Journal of Marketing, Financial Services & Management.

Indian Retail Segmentation

The Indian retail sector is going through the phase of tremendous transformation. The Indian retail sector is categorized into two segments, such as organized retail sector and unorganized retail sector with the later holding the larger share of the retail market. Though initially, the retail industry in India was mostly unorganized, however, with the change of tastes and preferences of the consumers, the industry is getting more popular these days and getting organized as well. The

Indian retail sector is highly fragmented with 96 per cent of its business being run by the unorganized retailers like the traditional family run stores and corner stores. The organized retail, however, is at a very nascent stage, though attempts are being made to increase its proportion bringing in a huge opportunity for prospective new players.

Indian Unorganised Retail Market

As per the National Accounts Statistics of India, the 'unorganised sector' includes units whose activity is not regulated by any statute or legal provision and/or those, which do not maintain regular accounts. Thus, unorganised retailing refers to the traditional formats of low-cost retailing, for example, the local kirana shops, owner managed general stores, paan/beedi shops, convenience stores, hand cart and pavement vendors, etc. Unorganised retailing is characterized as unstructured and high degree of fragmentation with street markets constitutes from peddlers, vegetable vendors, neighbourhood stores and consumer durable stores to manufacturer owned retail outlets.

Unorganised retail sector covers all those forms of trade, which sell an assortment of products and services ranging from fruits and vegetables to shoe repair. These products and services may be sold or offered out of a fixed or a mobile location and the number of people employed could range between 10-20 people. Thus, the neighbourhood baniya, the paanwala, the cobbler, the vegetable, fruit vendor, etc. would be termed as the unorganized sector. Traditionally, three factors have plagued the Indian retail industry.

Unorganized

India is known as the nation of shopkeepers, where vast majority of the retail stores are small "father and son" outlets. Traditionally, it is a family's livelihood, with their shop in the front and house at the back, while they run the retail business.

Fragmented

India has some 15 million retail outlets, however, a disturbing point is that 96 per cent of them are smaller than 500 square feet in area. This means that India per capita retailing space is about 2 square feet (compared to 16 square feet in the United States). India's per capita retailing space is, thus, the lowest in the world (Mathew *et al.*, 2008).

Rural bias

Nearly two-thirds of the stores are located in rural areas. Rural retail industry has typically two forms: "Haats" and "Melas". Haats are the weekly markets: serve groups of 10- 50 villages and sell day-to-day necessities. Melas are larger in size and more sophisticated in terms of the goods sold.

Opportunities and Challenges of Unorganized Retail Sector in India

Opportunities

- India's booming economy is a major source of opportunity. It is the third largest in the world in terms of purchasing power. India is the second fastest growing major economy in the world.
- Increasing instances of Double Incomes in most families coupled with the rise in spending power.
- Increased urbanization has led to higher customer density areas, thus, enabling retailers to use lesser number of stores to target the same number of customers. Aggregation of demand that occurs due to urbanization helps a retailer in reaping the economies of scale.
- With increased automobile penetration and an overall improvement in the transportation infrastructure, covering distances has become easier than before. Now a customer can travel miles to reach a particular shop, if he or she sees value in shopping from a particular location.

Indian unorganised retail sector challenges

- Lack of best practice in inventory management and supply chain management.
- Lack of standardization.
- Stiff competition from organized retail sector.
- Lack of knowledge, skills and training.
- Consumers shifting towards organized retail markets.
- Lack of government policies discouraging the unorganized retailers.

Indian Organised Retail Market

Organised retailing refers to trading activities undertaken by licensed retailers that is, those who are registered for sales tax, income tax, etc. These include the corporate-backed hypermarkets and retail chains, and also the privately owned large retail businesses. In other words, it is a network of similarly branded stores with an element of self-service.

Organised retail in India today holds only a fraction of the market share potential. In 2001-02, organized retail trade in India was worth [1] 11,228.7 billion. India's retail market is expected to grow to US$ 1.1 trillion by 2020 on the back of income growth, urbanization and attitudinal shifts. The organized retail sector, which is estimated to reach approx. 18-20 per cent of the total sector, by 2020, is growing at a high rate of 20-25 per cent per annum (S. Mishra, 2008).

Table 1.2: SWOT Analysis of Indian Organised Retail

Strength	Weakness
• Skyscrapers with perfect blend of shopping, eating and entertainment, in short "shoppertainment". • Developed in contemporary style, the flashy retail outlets promises just about everything under the sun, from foreign gizmos to the very desi. • Attractive destinations for civic and official meetings, hang out, reducing stress. • Procure goods directly from factories and farmers in case of lifestyle and food/beverages respectively. • Bouquet of value propositions like value for time, value for quality, value for experience, value for money. • An added service to residential developments, and thereby an added attraction.	• Most of the Indian organised retail outlets are owned neither by retailers nor by the developers but by financial investors. • Retail not accredited as an industry · Lack of adequate infrastructure including supply chain, parking facilities. • Unavailability and skyrocketing prices of prime catchment's areas. • Poor positioning and zoning of retail outlets. • Shortage of qualified human personnel in the area of facility management, creative firms, and design houses. • Lack of differentiated offerings i.e. same mix of shopping, foods and films.

Opportunities

- Tier II (Chennai, Pune, Hyderabad, etc. and Tier III cities(Indore, Jaipur, Goa,etc.) are still untapped
- Outsourcing from other developed retail market
- Progressive growth of aspirational consumer class, believing more in spending than savings
- Nation of youth (with median age 24 and 35 per cent of population below 14 years)
- Growing urbanisation and increase in purchasing power of consumers
- Abundant availability of skilled labour
- Emergence of India as the retail hub
- Low cost of operations

Threats

- Non-availability of adequate finance
- Market share of the unorganised retailers still ranges to 95 per cent and if not looked upon, might increase further.
- Disturbance in income strata of consumers greatly influence retail outlets growth
- Threat from online players, (even though internet penetration is low in India)
- Intrinsic complexities of retailing-rapid price changes, constant threat of product, etc.

Industry trends for retail sector indicate that organized retailing has major impact in controlling inflation because large organized retailers are able to buy directly from producers at most competitive prices. World Bank attributes the opening of the retail sector to FDI to be beneficial for India in terms of price and availability of products, as it would give a boost to food products, textiles and garments, leather products, etc., to benefit from large-scale procurement by international chains; in turn, creating jobs opportunities at various levels.

Food and Grocery Retail

Retail market for food and grocery is the largest of the different types of retail industries present in India. The All India food consumption was close to ₹ 9,000 billion, with the total urban consumption being around ₹ 3,300 billion during 2012-13. This means that aggregate revenues of large food players is currently only 5 per cent of the total Indian market, and around 15-20 per cent of total urban food consumption. During 2007-08, the food and grocery segment was valued at ₹ 7,920 billion and it enjoyed a dominant market share of 62 per cent in the total Indian retail sector; however, there was a completely opposite scenario in the organised retail segment (R.K. Srivastava, 2008).

Most food is sold in the local 'wet' market, vendors, roadside push cart sellers or tiny kirana stores. The share of an Indian household's spending on food is one of the highest in the world, with 48 per cent of income being spent on food and beverages.

Big Bazaar

Big Bazaar is an Indian retail store that operates as a chain of hypermarkets, discount department stores and grocery stores. The retail chain was founded by Kishore Biyani under his parent organisation Future Group, which is known for having a significant prominence in Indian retail and fashion sectors. Big Bazaar is also the parent chain of Food Bazaar, Fashion at Big Bazaar(abbreviated as *fbb*) and eZone, where at locations, it houses all under one roof, while it is sister chain of retail outlets like Brand Factory, Home Town, Central, eZone, etc. Founded in 2001, Big Bazaar is one of the largest hypermarkets chains of India, housing about 250+ stores in over 120 cities and towns across the country.

Local leaders and community members of Jammu joined Big Bazaar, its associates and hundreds of customers for the grand opening of Big Bazaar store on Oct, 21, 2016 at City Square Mall. The fashion sections have the best of fashions catering to men and women office wear, ethnic wear, casual wear, sportswear and also vast collection of kids wear. In the home and other sections, wide variety of home fashions from curtains, bedsheets, carpets, etc., and trendy collection of men and women footwear. Wide range of vegetables, fruits, staples, packaged food items, etc., will be available in the food section.

Various Competitors of Big bazaar

Easy Day is an Indian retail brand that runs chains of consumer retail department stores. The brand is wholly owned by Bharati Enterprises Limited and is operated by its subsidiary, Bharati Retail Limited, headquartered in New Delhi. The technical and management support for the brand is provided by Arkansas, United States based Wal-Mart Stores, Inc, which is the largest retailer in the world. Bharati Enterprises announced its foray into retail in February 2007 and the first store was opened in Punjab in April 2008. Easy Day presently has 220 stores across 13 states. The real estate, design and construction services to Easy Day are provided by Cedar Support

Services, an Indian retail company majority-owned by Bharati Enterprises, which also provides similar services to Bharati Walmart Private Limited

D-Mart seeks to be a one-stop shopping destination for the entire family, meeting all their daily household needs. A wide selection of home utility products is offered, including foods, toiletries, beauty products, garments, kitchenware, bed and bath linen, home appliances and much more. D'mart Exclusive is a part of internationally renowned Dolphin Group of Companies. D'mart Exclusive, a national player in luxury segment, offers exclusive lifestyle luxury decorative (art & décor pieces) from leading International design houses across the world.

Table 1.3: India's Retail-share of different categories (Billion US Dollars)

S. No.	Category	2006	2011	2016	CAGR (%)
1	Food and Grocery	217.00	325.00	425. 00	5.50
2	Restaurants & Food Junction	4.62.00	8.81	15.82	12.51
3	Apparel	25. 00	35. 00	50.21	7.50
4	Furniture & Furnishing	6.51	9.12	17.12	13.52
5	Healthcare & Fitness Services	0.43	1. 00	2.52	20. 00
6	Consumer Electronic & IT	16.51	22.70	42.84	13.51
7	Pharmacy	8. 00	13.94	23.44	11. 00
8	Jewellery, Watches, etc.	16.53	25.65	44.22	11.54
9	Beauty Services	0.63	1.34	3. 00	18. 00
10	Footwear	3.61	4.53	8.34	13. 00
11	Others	11. 00	23. 00	42.51	13.13
Total	US $ (Bn)	310. 00	420. 00	675. 00	7.53

Source: Computed from Technopak Advisors Pvt. Ltd data

Over the years, D'mart Exclusive has established itself as a name dedicated to bringing to India a wide collection offering range of aspirational and lifestyle collectibles. The extensive and varied collection that it offers include home décor and gifting solutions to the connoisseurs of all things of timeless beauty and impeccable quality ranging from Silverware, Handmade Crystal ware, Bronzeware, Handcrafted Marble Art Pieces, Fine Porcelain figurines, Limited Edition pieces, Furniture Accessories and gifts Items.

D'mart Exclusive is the pioneer in introducing a collection of Worldwide Limited Edition representations of Indian deities in 92.5 per cent sterling silver and fine Italian porcelain. Conceptualized and designed in India, these masterpieces have been hand-crafted to perfection by master craftsmen from some of the most sought after brands in the world. Headquartered in New Delhi, D'mart Exclusive has a Nationwide Franchisee network, with 14 premium showrooms and boutiques in New Delhi, Mumbai, Chennai, Kolkata, Bengaluru, Hyderabad, Nagpur, Pune, Shridi, Chennai, Vishakapatnam in India, alongwith international representations in Amsterdam, New Jersey and San Francisco.

From the above listed table, it can be affirmed that Food and Grocery constitutes the bulk of Indian retailing and its share is estimated at US$325 Bn in 2011 (69 per cent of the overall retail) and is expected to grow to US$ 425 Bn by 2016 (63 per cent of the overall retail), @ CAGR of 5.50 per cent.

❑❑❑

Chapter 2

Theory of Customer Satisfaction in Retailing

As the retail industry is getting a lot of market share in India, in the background of changing advertisement trends, understanding customer perceptions, building relationship, marketing of products and retaining customers has been identified as major source of competitive advantage. Businesses around the world are becoming increasingly competitive day-by-day. Therefore in order to generate more customers and retain existing ones, companies engage in various forms of marketing activities. Among all those activities, "Customer Satisfaction" has key role.

Customer satisfaction

Customer satisfaction is the measure of how the needs and responses are collaborated and delivered to excel customer expectation. It can only be attained, if the customer has an overall good relationship with the supplier. In today's competitive business marketplace, customer satisfaction is an important performance exponent and basic differentiator of business strategies. Hence, the more is customer satisfaction, more is the business and the bonding with customer.

Customer satisfaction is a part of customer's experience that exposes a supplier's behaviour on customer's expectation. It also depends on how efficiently, it is managed and how promptly services are provided. This satisfaction could be related to various business aspects like marketing, product manufacturing, engineering, quality of products and services, responses customer's problems and queries, completion of project, post delivery services, complaint management, etc.

Customer satisfaction is the overall essence of the impression about the supplier by the customers. This impression, which a customer makes, regarding supplier is the sum total of all the process he goes through, right from communicating supplier before doing any marketing to post delivery options and services and managing queries or complaints post delivery. During this process, the customer comes across working environment of various departments and the type of strategies involved in the organization. This helps the customer to make strong opinion about the supplier, which finally results in satisfaction or dissatisfaction.

Consumer buying behaviour

Consumer buying behaviour is the result of the attitudes, preferences, intentions and decisions made by consumers in a market place before buying a product. Consumer behaviour is very important to the marketers because it enables them to understand and predict buying behaviour of consumers in the market place. It is concerned not only with what consumers buy, but also with why they buy it, when and where, and how they buy it and how often they buy it and also, how they consume it and dispose it. According to Professor Theodore Levitt of the Harvard Business School, the study of Consumer Behaviour is one of the most important in business education, because the purpose of a business is to create and keep customers.

Consumer behaviour involves the study of how people—either individually or in groups-acquire, use, experience, discard, and make decisions about goods, services, or even lifestyle practices. There are intra-personal, inter-personal and many other environmental variables that affect consumer's decision-making process. The inter-personal influences include family, social class, reference group, cultural and financial status of an individual. The intra-personal variable includes motivation, perception, learning, attitudes and personality of the buyer.

Competition with regard to service

Competition in the commercial world means firms striving to win customers, with innovative firms bringing ideas to market, successful firms thriving, and unsuccessful ones making an exit. When markets are competitive, consumers will be offered variety and choice, with firms striving to win customers on the basis of service, quality, price and innovation. In such circumstances, consumers can feel confident in exercising choice and competition is strengthened.

In a number of different retail categories, price has always been seen as a key driver to "competitiveness .Competition can benefit consumers in different ways . Competition may lead to greater product variety, higher product quality, and greater innovation, which drives productivity growth and helps in lifting living standards (Hotelling 1929; Aghion *et al.*, 2005; Shapiro, 2012). When there, is little or no competition, consumers are made worse off.

Review of Studies related to the Customer Satisfaction

The review of literature guides for better understanding of the methodology, used by earlier studies and also, the limitations of various available estimation procedures, database, logical interpretation and understanding conflicting results.

Oliver (1981) described in his study that customer satisfaction is the part of marketing and play important role in the market. In any organization, satisfaction of customer is more important, because if your customer is satisfied with your services or products, your position will be good in the market. In old times customer satisfaction was not too much important and people were not focused on quality. But now-a-days competition is tough and customer is aware of all the products and companies due to education and learning environment and this is the reason that every business is concerned to fulfill customer's need and wants properly.

Westbrook and Reilly (1983) stated customer satisfaction as "an emotional response to the experiences provided by, associated with particular 32 products or services purchased, retail outlets, or even molar patterns of behaviour, such as shopping and buyer behaviour, as well as the overall market place".

Zeithaml (1985) stated that customer satisfaction is a scale with multiple items good validity and reliability, many researchers use this instrument to understand and find the service expectation and consumer's perception for improving the service quality. The expectation of a customer is an important factor. The more service quality is near to customer expectation, there will be more satisfaction and vice versa.

Tse and Wilton (1988) stated that "the consumer's response to the evaluation of the perceived discrepancy between prior expectations (or some other norm of performance) and the actual performance of the product/service is perceived after its consumption".

Berry and Parasuraman (1991) stated that the customers' satisfaction is influenced by the availability of customer services, the provision of quality customer service and has become a major concern of all businesses. Customer satisfaction is typically defined as a post consumption evaluative judgement, concerning a specific product or service. It is the result of an evaluative process that contrasts pre-purchase expectations with perceptions of performance during and after the consumption experience.

Oliver (1993) suggested that customer satisfaction is the core philosophy of marketing strategy of any organization and plays a key role in an organization success. In fact, customer satisfaction is the core principle of the modern tool of CRM being used by marketers to attract and retain customers.

Anderson *et al.* (1994) stated that there is a positive relationship between customer satisfaction and economic profit for the company. Arguably, customer satisfaction is an important component in order for the company to be profitable.

Anton (1996) stated "customer satisfaction as a state of mind in which the customer's needs, wants and expectations throughout the product or service life have been met or exceeded, resulting in subsequent repurchase and loyalty".

Fornell *et al.* (1996) defined customer satisfaction with three manifest variables: customer satisfaction as a whole, as an estimation how expectations were fulfilled, and as comparison with the ideal service or product. They believe this choice is consistent with the orientation of American Consumer Satisfaction Index (ACSI).

Woodruff and Gardian (1996) defined "Satisfaction as the evaluation or feeling that results from the disconfirmation process. It is not the comparison itself (i.e., the disconfirmation process), but it is the customer's response to the comparison. Satisfaction has an emotional component."

Thorsten and Alexander (1997) found customer satisfaction with the product and services of company as the strategic factor for competitive advantage. In the context of relationship marketing, customer satisfaction is the way that leads to long-term customer retention because unsatisfied customers have very high switching rate.

Vavra(1997) defined customer satisfaction as a satisfactory post-purchase experience with a product or service given an existing purchase expectation.

Zairi (2000) depicted that feeling of accomplishment of inner desires is called satisfaction. Customer satisfaction has direct effect on customer loyalty (Mittal & Lassar, 1998). If product or service fulfills the needs and demand of customer, he will become satisfied and will be converted to loyal customer and thus, will add in customer equity of company. Company profitability is not only depicted in its balance sheet but it is also measured on the basis of its sound customer base and lifetime value that customers deliver to company.

Ranaweera and Prabhu (2003) stated that the more satisfied customers are, the greater is their retention, the positive word of mouth generated through them and the financial benefits to the firms, who serve them. It is not surprising, therefore, that the fundamental aim of firms is to seek to manage and increase customer satisfaction at least in this era of competitive global marketing.

Zeithaml *et al.* (2003) defined customer satisfaction as customer's evaluation of a product or service in terms of whether the product or service has met his needs or expectations. Failure to meet needs results in dissatisfaction, or a poor perception of the service quality. Satisfaction can be acknowledged in various senses depending on what needs the customer had before the service; it ranges from feelings of fulfillment, contentment, pleasure, delight, relief, and ambivalence. Although it tends to be measured as a static quantity, it is dynamic and evolves over time being influenced by a variety of factors.

Zeithaml & Bitner (2003) stated that "Satisfaction can be regarded as a broad principle; support good quality is a component of satisfaction".

Schiffman and Kanuk (2004) defined customer satisfaction as "The individual's perception of the performance of the product or service in relation to his or her expectations".

Kotler and Keller (2006) defined satisfaction as a person's feeling of pleasure or disappointment, which resulted from comparing a product's perceived performance or outcome against his/her expectations.

Kurniawan (2010) stated that branding is one of the factors, which has been identified repeatedly customer choice in the selected product. Branding has impact on purchase decision of consumers. The more a brand has succeeded in its earlier period, the more value for its brand can be generated.

Mohsam, *et al.* (2011) observed that the issue of customer satisfaction has been gaining increasing attention among researchers and practitioners as a fundamental tool in financial institutions for enhancing organizational performance and profitability.

Rajasekara and Saravanan (2014) studied consumer satisfaction on fast moving consumer goods, which focused on the attitudes, values, beliefs and perception of the consumers with regards to the consumption of fast moving consumer goods. To study the brand preference of fast moving consumer goods, the preference and opinion of consumer towards popular brands have been taken into consideration.

Review of Studies related to the Consumer Buying Behaviour

Krugman (1965) defined the consumer buying behaviour as the actions, thought process and perceived outcome, in collaboration with environmental factors, during the course of making a decision, which could result in a purchase. Some of the factors that are significant in consumer behaviour are external environment, demographics and personal characteristics, which are influenced by the consumer's beliefs, values and attitudes.

Engel, *et al.* (1968) defined consumer behaviour as "those acts of individuals directly involved in obtaining, using, and disposing off economic goods and services, including the decision processes that precede and determine these acts". Simple observation provides limited insight into the complex nature of consumer choice and researchers have increasingly sought the more sophisticated concepts and methods of investigation provided by behavioural sciences in order to understand, predict, and possibly control consumer behaviour more effectively. Psychology, social psychology, and sociology are the disciplines most widely employed in this endeavour, which has become a substantial academic industry in its own right.

Rees (1992) investigated the factors influencing the consumer's choice of food and found as flavour, texture, appearance, advertising, a reduction in traditional cooking, fragmentation of family means and an increase in 'snacking', etc. Demographic and household role changes and the introduction of microwave ovens had produced changes in eating habits. Vigorous sale of chilled and other prepared foods was related to the large numbers of working wives and single people, who require value convenience. Development in retailing with concentration of 80 per cent of food sales in supermarkets was also considered to be important. Consumers were responding to messages about safety and healthy eating. They were concerned about the way in which food was produced and want safe, 'natural', high quality food at an appropriate price.

Engel *et al.* (1993) stated that consumer buying behaviour is directly related to the consumption of products and services, which are triggered by the decision-making process before and after. He also expressed that the purchase behaviour of the narrow explanation refers to the buying behaviour of individuals substitutes the exchange of goods or services in money or money.

Assael and Henry (1994) explained that consumer behaviour as the process of perceiving and evaluating different components of a purchase. Furthermore, he explains that there are two broad influences that decide consumer choice. At first, the individual consumer's needs, consciousness of brand characteristics and awareness to alternatives are of significance in the matter of decision-making. Other factors that influence the choice of brand are the consumer's demographics, lifestyle and personal characteristics. Secondly, the environment has importance in consumer behaviour, which consists of culture, social class and reference groups..

Ha¨ubl, G. (2000) stated that the consumer behaviour about decision-making is difficult to define and is a system of short cuts and rule of thumb, which is unpredictable. The short cuts in decision-making vary from person to person and focusing on the past experience of consumers; we can predict the future trends by bringing profitable products and services into the market.Keys, T. P., (2011), stated that "We're not aware of changing our minds even when we do change our minds. And most people, after they change their minds, re-construct their past opinion - they believe they always thought that."

Srinivasan (2000) revealed that consumer with higher educational level was found to consume more processed products. The quantities of processed fruit and vegetable products were consumed more in high income group. The tolerance limit of price increase identified was less than 5 per cent, any price change above this limit, would result in discontinuance of the use of the processed product. Consumers preferred processed products because of convenience of ready-to-eat form.

Blackwell *et al.* (2001) defined consumer behaviour as a summation of acquisition, consumption and disposal of products or services.

However, such definition falls short of the continuity of the processes. Based on this loophole, Arnoud *et al.* (2004) further proposed the circle of consumption that 16 recognize purchasing processes as a loop, comprising acquisition of goods and services, consumption, as well as disposal of used goods. As far as the consumer decision process model is concerned, consumers need to go through seven steps before reaching their final decisions. These seven steps include need recognition, search for information, pre-purchase, evaluation, purchase, consumption, post-consumption evaluation and divestment.

Hawkins *et al.* (2004) stated various factors affecting consumer behaviour for buying, such as demographic and social influences (family and household), group influence, impact of advertising and internal influences (learning, perception, attitude, etc.). They also elucidated the topics, such as types of consumer decisions, purchase involvement and product involvement. They emphasized on information search process and various ways for providing relevant information to the consumers. They studied individual judgment and proposed that the ability of an individual to distinguish between similar stimuli is called sensory discrimination, which could involve many variables related to individual preferences.

Porter (2004) pointed out that firms can create value by providing lower price or unique offers to the customers so as to excel their competitive advantages over the others. Stage four refers to the purchase decisions made by the consumers after evaluating the offers from different retailers.

Kotler *et al.* (2005) explained that the pre-purchase evaluation that consumers compare between different products and brands to make a purchasing decision. In this stage, consumers pay particular attention to the attributes, which are most relevant to their needs. Attributes like quantity, size, quality and price are commonly used to judge a brand by customers. Any changes in these attributes can affect consumers' decisions on brand or product choices.

Blackwell *et al.* (2006) stated that there are two phases contributing to the decision-making processes, including retailer and in-store selection. Retailer selection is made by judging, which retailers to buy after investigating the attributes from the previous stage, whereas in-store selection is affected by the selling skills of salesperson, visual displays inside the shops, as well as point-of-purchase advertising.

Solomon *et al.* (2006) stated that along the same lines the complexity of consumer behaviour with regards to choice comes to our attention, when we see it under the light of consumer behaviour motivations. Consumers' motivations are frequently originated in underlying aspiration values. Further, consumers are not usually willing to communicate these underlying desires or often are not aware that their product evaluation behaviour is an extension of whom they are and who they want to be.

Kotler *et al.* (2007) stated that usually a buyer passes through five stages to reach his buying decision. First the buyer notices the difference between his current state and ideal state, and recognizes his want and need for something. A need can also be aroused by external stimuli. He/She starts searching the information for his/her desired product through different channels like family, friends, advertisement or mass media. After sufficient information is gathered, the buyer processes the information to evaluate the alternatives brands in the choice set. Finally, he purchases the product which he assumes to be the best for him. After purchasing the product, the buyer will take further action to the marketer based on his satisfaction or dissatisfaction.

East *et al.*(2008) stated that purchase behaviour have been accepted that most repeat purchase decisions are done as a habit and often this range of decision-making from rational to automatic is related to the degree of involvement.

Kapferer (2008) pointed out that the brand is the ultimate personality, which discards typical, dissonant elements in disagreement with its personality. A brand provides consumers a sense of stability and consistency. Along those lines, a brand is less flexible than its product categories. Once the personality image of the brand is created in the minds of consumers, it is very difficult to change. Thus, therein lays the importance of understanding consumers' rejection of products and brands.

Saravanan (2010) revealed in his study that the modern market is highly competitive and traditional. The prominence gained by an individual consumer's decision-making in marketing compels the marketers to look at and organize the component of the marketing mix through the customer's eyes. Hence, consumer behaviour research has come into existence. In the present era, women play a vital role in all fields and women play a major role in taking purchase decisions for non-durables.

Arutselvi (2012), studied preference towards various types of Britannia biscuits in Kanchipuram towns. The consumer behaviour varies from brand to brand on the basis of quality, quantity, price, taste, advertisement, etc. It is concluded that the market study on biscuits at Kanchipuram town has helped to know the status of biscuits. It has revealed the requirements of the taste of the consumer of biscuits. Britannia biscuits are having a good market share in Kanchipuram town.

Kuester (2012) stated that the consumer behaviour is the study of individuals, groups, or organizations and the processes they use to select, secure, and dispose off products, services, experiences, or ideas to satisfy needs and the impacts that these processes have on the consumer and society.

Jegan and Sudalaiyandi (2013) studied on consumer buying behaviour towards various types of Sunfeast (ITC) biscuits in Kovilpatti to know the consumers' preferable taste, awareness about various brands, about the choice and their frequency of preference, satisfaction of Sunfeast biscuit. The result of the study shows that Sunfeast biscuits have a good market share in Kovilpatti city.

Review of Studies related to the Business competition

Porter (1980) stated that much competitive information is bounded by the assumptions that managers have with respect to their industry and these assumptions may lead to *blind spots*. The effect of such blind spots may cause the strategist not to recognize the significance of events, interpret them inappropriately, or see them only slowly.

Day (1998) stated that beneath a competitor's marketplace strategy lie the organization and the functional operations and processes that make the strategy possible. If the competitor is rational, then its marketplace strategy will have been built around those functions and activities, where it is competitively advantaged versus competitors.

Karni *et al.* (1998) studied and stated that the concept of interfirm rivalry extends beyond the product/market level. Competition can also occur as firms use related resources to bear on individual product/

market level rivalry. The theory of multi-market competition describes those situations, in which firms compete against each other inmultiple markets.

Luo (1999) investigated environmental factors, such as complexity, dynamism and munificence, influencing small firms' performance. Based on an analysis of survey data collected from SMEs, Luo confirmed that environmental characteristics have a significant influence on SMEs' strategic orientations.

□□□

Chapter 3

Fundamentals of Retailing

The exercises engaged with the offering of merchandise to extreme customers for individual or family unit utilization is called retailing. The word 'Retail' is gotten from a French word with the prefix re and the action word posterior signifying "to cut once more". Clearly, retail exchange is one that cuts off littler bits from substantial chunks of merchandise. It is a procedure through which products are transported to definite buyers. As such, retailing comprises of the exercises associated with offering specifically to a definitive shopper for individual, non-business use. It grasps the direct-to-client deals exercises of the maker, regardless of whether through his very own stores by house-to-house peddling or via mail-arrange business. A retailer is a dealer or at times a specialist or a business endeavour, whose principle business is offering straightforwardly to extreme customers for non-business use. He performs many advertising exercises, for example, purchasing, moving, evaluating, chance exchanging, and creating data about client's needs. A retailer may offer rarely to mechanical clients, yet these are discount exchanges, not retail deals. In the event that more than one portion of the measure of volume of business originates from deals to extreme buyers, i.e. deals at retail, he is named a retailer. Retailing happens in all promoting channels for shopper items.

The retailer is a middle person in the advertising channel since he is both advertiser and client, who pitches to the last man to expend. He is an expert, who keeps up contact with the purchaser and the maker; and is an imperative associating join in an intricate component of showcasing. In spite of the fact that makers may offer straightforwardly to shoppers, such strategy for conveying merchandise to extreme clients is awkward, costly and tedious, when contrasted with the activity performed by a master in the line. Along these lines, as often as possible, the makers rely upon the retailers to pitch their items to a definitive customer. The retailer, who can give proper civilities without an exorbitant development in costs of merchandise is compensated by bigger or increasingly faithful support.

In view of the intrigued clients, space accessible, rivalry at the place retailers pick distinctive organizations and models for moving their items. Prior the retail outlets accessible used to be the conventional kiryana stores, yet with the advancement in retailing, sorted out retailing is expanding at an exponential rate. The following is the point by point portrayal of various sorts of the retail organizes. It is in retailing that extremely uncommon changes have happened amid the most recent two decades. A few establishments have vanished while more up-to-date ones have been included. This procedure of erasure/ expansion still proceeds in more current structures. There are large scale retailing shops together with little units, both working all the while. They have from sellers and vendors, who have no changeless place, to efficient, settled retail shops like chain stores, departmental stores, and so on. The foundations carrying on the retail business can be named under.

Formats in Retailing

Retailers extend in size, shape little, autonomous, proprietor worked shops like the nearby flower specialist, cleaners or hair stylist, to national and global monster classification executioners. The fundamental characterization done is based on store based retailers and non-store retailers. The store based retailers can be additionally arranged based on the stock that they offer, or by the way of possession.

Types of Retail Stores

- Instore-Retailing
- Non-store Retailing
- Franchising

Instore-Retailing

1. Department Stores
2. Super Markets
3. Markdown Houses
4. Chain Stores
5. Various Shops

Non-store Retailing

1. Direct Selling
2. Telemarketing
3. Online Retailing

Department Stores

These are extensive scale retail locations moving under one rooftop and one control an assortment of products isolated into various divisions, every one of which has some expertise in an individual stock. Talk is of the feeling that a retail establishment is a retail shop taking care of a few classes of products including quick moving shopper merchandise, each class being isolated from others in the executives, book-keeping and area. It is seen by Clarke as that kind of retail organization, which handle a wide verity of stock under one rooftop which the stock gathered into all around characterized offices, which is midway controlled and which cooks essentially to ladies customers. Hence, a retail chain is a retailing speciality unit that handles a wide assortment of shopping and claim to fame products and is composed into isolated divisions for reasons for deals advancement, book-keeping control and store activity. Ongoing patterns are to include divisions for car, recreational and sports gear, too an administrations, for example, protection, travel guidance and salary assess planning. Retail chains are particular in that they as a rule are situated towards administration. They are generally strip malls.

These stores might be arranged either as per proprietorship or pay gatherings to which they advance.

a) Based on possession, these are:
   i) The autonomous;
   ii) The proprietorship gathering; and
   iii) Chain retail establishments.

   Autonomous stores are claimed by a money related premium, which does not possess other comparable stores. Ownership aggregate stores are those stores, which were previously needy, yet now have been consolidated.

b) Based on pay gatherings: These stores take into account the center and high salary gatherings. They ordinarily handle great quality stock and offer most extreme support of the clients. Different stores oblige the necessities of the lower pay bunch individuals.

c) Sometimes, there is likewise to be discovered, what are called rented retail establishments. Despite the fact that it appears to most clients that all offices in a retail establishment are possessed and worked by the store, that isn't generally the situation. The activities of specific divisions are in some cases swung over to leases, and such offices are called rented offices.

Trademark Features of Department Stores

The main highlights of these stores are:

These are incorporated stores performing activities notwith standing other retail locations, for example, wholesaling :

i) Goods are separated into various classes with various areas and the board inside the store itself.

ii) These stores are recognized by the idea of products they self and not by the assortments they keep for instance, medication and assortment stores convey a wide assortment of merchandise.

iii) The store is an evenly incorporated organization. It unites under one rooftop, a scope of stock contributions equivalent to the joined contributions of numerous stores represent considerable authority in single or less stock lines.

Area of Department Stores

The achievement of a retail chain depends much on its area, accessibility of room, the zone and network to be served and capacity to draw in clients are the vital elements to be considered before building up a store at a specific place. Unique consideration ought to be given to settlement in order to permit each conceivable diversion offices. Extensive space ought to likewise be designated for showroom showing stores stock.

Benefits of Department Stores

1. Huge retail establishments purchase in vast amounts and get uncommon concession or rebate in their buys. A large number of them buy coordinate frame fabricates and consequently, go between charges are disposed off.
2. Retail establishments are in a situation to pay money on all or the majority of their buys and this gives them an extra favourable

position of getting quality products at less expensive rates and in the meantime stocking the most recent style and trends.

3. Clients can do every one of their buys under one rooftop and it bids to individuals of varying backgrounds. The association is too substantial to even consider providing master supervision of different divisions for the appropriation of a liberal credit and conveyance benefit for huge scale promoting.
4. At the point when clients enter the store to manage one office, they are as often as possible prompted by the promotion, which the presentation of products offers to make buys in different divisions too.
5. The expense of working together is high because of overwhelming overhead costs.
6. On account of their area in a focal shopping territory, they are of very little preferred standpoint to people in general, since merchandise required at short notice are constantly obtained from the closest dealers.
7. There is absence of individual touch and individual supervision, which is to be found in single line.
8. At the point when contracted tirelessness is substituted for the persistence of proprietorship, misfortune and breaks are probably going to happen.
9. Numerous clients misuse the liberal administrations expanded and exploit the strategy of the 'clients is in every case right'.
10. The kind of charisma found in numerous stores is exceptionally poor in light of low installments and absence of supervision.

Future of Department Stores

Nothing definite can be predicted whether these department organizations will continue to carry on and will progress in face of overgrowing competition of chain stores, mail order business and other smaller independent dealers. Stores with overhead burdens, which cannot be reduced, may have to go, but the department stores, as an institution are bound to go on with a future. The department store which is properly equipped in plant, stock and personnel to carry on a reasonable sales volume and then does a better job in giving values and services, then its competitors are entitled to, and will receive its business profit. Department stores are now opening branches in many new areas and making concerted efforts to meet new competition. They have been modernized, re-decorated and better services are being developed; and they are being converted to self-service.

Supermarkets

These are large, self-service stores that carry a broad and complete line of food and non-food products. They have central check out facilities. Kotler defines supermarket as 'a departmentalised retail establishment having four basic departments viz. self-service grocery, meat, produce and diary plus other household departments, and doing a maximum business. It may be entirely owner operated or have some of the departments leased on a concession basis.

Characteristic Features of Supermarkets

Chief characteristic features of supermarkets include the following:

i) They are usually located in or near primary or secondary shopping areas but always in a place, where parking facilities are available.

ii) They use mass displays of merchandise.

iii) They normally operate as cash and carry store.

iv) They make their appeal on the basis of low price, wide selection of merchandise, nationally advertised brands and convenient parking.

v) They operate largely on a self-service basis with a minimum number of customer services.

Supermarkets came into existence during the depression in U.S.A. At that time they sold only food products, and their principal attraction was the low price of their merchandise. As supermarkets increased in number day-by-day, they also expanded into other lines of merchandise.

Advantages of Supermarkets

i) Supermarkets have the advantage of convenient shopping, permitting the buyer to purchase all his requirements at one place.

ii) Supermarkets also stock a wide variety of items.

iii) These markets can sell at low prices because of their limited service feature, combined with large buying power and the willingness to take low per cent of profit margins.

iv) Shopping time is considerably reduced.

Limitations of Supermarkets

i) The large and extensive area required for a supermarket is not available cheaply in important places.

ii) The products which require explanation for their proper use cannot be dealt in through the supermarkets.

iii) Customer services are practically absent.

iv) Another limitation of the supermarket is the exorbitantly high administrative expenses.

Discount Houses

These are large stores, freely open to the public and advertising widely. They are self-service and general merchandising stores. They carry a wide assortment of products of well known brands, appliances, housewares, home furnishings, sporting goods, clothing, toy and automotive services. They complete on low price basis and operate on a relatively low mark-up and a minimum number of customer service. They range from small open showroom to catalogue type order offices to full line limited service, and promotional stores. They buy their merchandise stocks both from wholesale distributors and directly from the manufacturers.

Chain Stores or Multiple Shops : A chain store system consists of four or more stores, which carry the same kind of merchandise are centrally owned and managed and usually are supplied from one or more central warehouses. A chain store is one of the retail units in chain store system. Chains have been interpreted as a group of two or more reasonably similar stores in the same kind or field of business under one ownership and management, merchandised wholly or largely from central merchandising headquarters and supplied from the manufacturer or orders placed by the central buyers. In Europe, this system is called as Multiple Shops and the American call it as "Chain Stores". Under the multiple or chain shop arrangement, the main idea is to approach the customers and not to draw the customers, as it is practiced in the case of department store. In order to draw more customers, attempts are made to open a large number of shops in the same city at different places. In India, apt example for this retail system are offered by 'Bata Shoes', 'Usha Sewing Machines', etc., such multiple shops have 'centralized buying with decentralized selling". Fundamentally, they specialize in one product but with all its varieties or models.

Online Retailing

When a firm uses its website to offer products for sale and then individuals or organisations use their computers to make purchases from this company, the parties have engaged in electronic transactions (also called online selling or internet marketing). Many electronic transactions involve two businesses, which focuses on sales by firms to ultimate consumers. Thus online retailing is one which consists of electronic transactions in which the purchaser is an ultimate consumer. Online retailing is being carried out only by a rapidly increasing number of new firms, such as Busy.com, Pets Mart and CD Now.com. Some websites feature broad assortments, especially those launched by general merchandise retailers, such as Wal-mart and Target. Some Internet only firms, notably Amazon.com are using various methods to broaden their offerings. Whatever their differences, e-retailers are likely to share an attribute. They are unprofitable or best, barely profitable. Of course, there are substantial costs in establishing an online operation. Aggressive efforts to attract shoppers and retain customers through extensive advertising and low prices are also expensive. The substantial losses racked by online enterprises used to be accepted, perhaps even encouraged by investors and analysts. The rationale was that all available funds should be used to gain a foothold in this growing market. Despite these challenges, online retailing is expected to grow, rapidly and significantly for the foreseeable future. Online sales represented about 1% of retail spending in 2005, but one research firm estimates that consumer purchases on the Internet with triple by the year 2010. Which product categories are consumers most likely to buy on the Internet in the future? Consumers' shopping intentions in 2005, placed the following goods and services at the top of the list: books, music and videos, computer hardware and software, travel and apparel. Of course, given that change on the Internet occurs, these categories soon may be surpassed by others - perhaps groceries, toys, health and beauty aids, auto parts or pet supplies.

Automatic Vending

The sale of products through a machine with no personal contact between buyer and seller is called automatic vending. The appeal of automatic vending is convenient purchase. Products sold by automatic vending are usually well-known pre-sold brands with a high rate of turnover. The large majority of automatic vending sales comes from the "4 c's" : cold drinks, coffee, candy and cigarettes. Vending machines can expand a firm's market by reaching customers where and when they cannot come to a store. Thus, vending equipment is found almost everywhere, particularly in schools, workplaces and public facilities. Automatic vending has high operating costs because of the need to replenish inventories frequently. The machines also require maintenance and repairs. The outlook for automatic vending is uncertain. The difficulties mentioned above may hinder future growth. Further, occasional vending-related scams may scare some entrepreneurs away from this business. Vending innovations give reason for some optimism. Debit cards that can be used at vending machines are becoming more common. When this card is inserted into the machine, the purchase amount is deducted from the credit balance. Technological advances also allow operators to monitor vending machines from a distance, thereby reducing the number of out-of-stock or out-of-order machines.

Direct Marketing

There is no consumers on the exact nature of direct marketing. In effect, it comprises all types of non-store retailing other than direct selling, telemarketing, automatic vending and online retailing. In the context of retailing, it has been defined as direct marketing as using print or broadcast advertising to contact consumers, who in turn, buy products without visiting a retail store. Direct marketers contact consumers through one or more of the following media: radio, TV, newspapers, magazines, catalogues and mailing (direct mail). Consumer order by telephone or mail. Direct marketers can be classified as either general - merchandise firms, which offer a variety of product lines, or speciality firms, which carry - only one or two lines, such as books or fresh fruit. Under the broad definition, the many forms of direct marketing include:

- Direct mail - in which firms mail letters, brochures and even product samples to consumers, and ask them to purchase by mail or telephone.
- Catalogue retailing - in which companies mail catalogues to consumers or make them available at retail stores.
- Televised shopping - in which various categories of products are promoted on dedicated TV channels and through infomercials, which are TV commercials that run for 30 minutes or even longer on an entertainment channel. On the plus side, direct marketing provides shopping convenience. In addition, direct marketers enjoy comparatively low operating expenses because they do not have the overhead of physical stores. Direct marketing has drawbacks. Consumers must place orders without seeing or touching the actual merchandise. To off-set this, direct marketers must offer liberal return policies. Furthermore, catalogues and to some extent, direct mail pieces are costly and must be prepared long before they are issued. Price changes and new products can be announced only through supplementary catalogues or brochures. Direct marketing's future is difficult to forecast, given the rise of the Internet. The issue is whether or not firms relying on direct marketing can achieve and sustain a differential advantage in a growing competition with online enterprises.

Franchising

A franchising operation is legal contractual relationship between a franchiser (the company offering the franchise) and the franchisee (the individual, who will own the business). The terms and conditions of the contract vary widely but usually the franchiser offers to maintain a continuing interest in the business of the franchisee in such areas as the site selection, location, management, training, financing, marketing, record-keeping and promotion. He also offers the use of a trade name, store motif standardized operating procedure and a prescribed territory. In return the franchisee agrees to operate under conditions set forth by the franchiser. For the manufacturers, the franchising is beneficial in these directions:

i) It allows them to conserve capital.

ii) The distribution system is established in the shortest possible time,

iii) Marketing costs are lowest, and

iv) Expenses of fixed overhead, such as administrative expenses of the personnel of the company owned units are cut down substantially. Franchising exists in such products as soft drinks, automobiles and parts, business services, dry cleaning, etc. The franchisee should also:

- Make reference check from the financial institutions.
- Make inquiries about the product, its quality, appeal, exclusiveness, competitiveness and effectiveness in bringing in repeat customers.
- Have enough capital to buy the franchise,
- Be capable of taking supervision work.
- Consult the professionals and seek their guidance in legal matters,
- Take risks and invest sufficient time

Theory of Retailing

Explaining How different retail formats Emerge and Mature.

At different times, different retail formats have been popular. Strong retail formats have become marginal and new retail formats have often emerged to dominate the retailing scene. Wheel of Retailing explains how different retail formats emerge, mature and are then replaced by another format.

Wheel of Retailing

The wheel of retailing concept was introduced by McNair from Harvard University and it is considered to be more an observation than a theory. No matter from which point of view we look at this concept, the idea itself intends to describe how the retail institutions transform during their evolutionary lifecycles.

Establishing and Penetrating in the Market

The theory pays attention to the new retailers, which often enter the market place with low prices as well as low profit margins and sometimes low status. The low prices are usually the result of some innovative cost-cutting procedures. Sooner than later, these innovative cost-cutting procedures will most probably attract competitors, if the entry barriers are not high enough.

Example - When book stores like Barnes and Nobles started, they started by a unique model of bringing all the books under one roof with amazing discounts. This was the first stage for them in the wheel of retailing.

Expanding in the Market

During the time and while they gain more experience of the market, these retailer businesses strive to enlarge their customer base with the purpose of increasing sales, gaining a higher profit margin as well as acquiring a significant market share. They can attempt to increase their customer base through different mechanisms, most of the times by attempting to target a different segment of customers. Since these retailers initially enter the market with low prices, their main target is represented by low income population.

Example - Barnes and Nobles then started expanding the market with more and more stores so that they increased their sale, their brand value and ultimately, their margins started increasing as well.

Stabilized business model attracting margins

In this stage of the Wheel of retailing, the company is already in an established position and hence, the rates are enough to get a decent margin. Because of the margin it is getting, the company keeps expanding moderately and increases its reach to attract as well as retain more customers. This is the most profitable stage for any retailing organization. However, when there is more demand than supply, it always opens up a niche. And generally, at this stage itself, other competitors start preparing or analyzing the market and think on how to penetrate this market.

The entry by another retail competitor who challenges, and then brings down the original

By adding higher qualitative products to the market or by providing additional services, or by simply moving to a better market location, the retail businesses then target another segment. Therefore, their operations and facilities increase and become more expensive, as they might require extra labour, extra expertise, extra warehouse, etc.

Ultimately, these retail businesses might emerge as a high cost price service retailer.

This is done in order to recover its fixed costs quickly and have an early breakeven so that it can start generating some profit. Overall, in this stage, the cost is high for the original retailer. And hence, its prices are higher. Therefore, this presents an opportunity for another retailer who can then enter in the market again in stage 1, with an objective to penetrate the market. New retail businesses then enter in the market and attempt to follow the same steps as the previous one before.

Based on this, the wheel of retailing concept is seen as a cycle and an evolutionary theory and it represents one of the theories of structural change in retailing. As the theory involves the beginning represented by one state and return to the same state after some time in the future, the theory is perceived as being cyclical.

Challenges in the Wheel of Retailing Concept

There are a few limitations concerning this theory in terms of focusing only on margins and prices, when there are also other variables affect the evaluation of the retail sector, such as environmental changes and competition. Moreover, the discussed theory cannot apply to all retail businesses, as there are also businesses, such as boutiques, vending machines and convenience stores, which are being operated with a high margin basis from the entry phase.

□□□

Chapter 4

Role of Customer Satisfaction in Retailing

Consumer satisfaction is a standout amongst the most focal markers, when estimating and examining organization's prosperity probability now and in future. The normal business loses 10-30 per cent of its clients every year, except they regularly don't know, which clients they have lost, when they were lost, why they were lost, or how much deals income and benefit this client rot has cost them. (Nigel Hill, Handbook of Consumer Satisfaction and Satisfaction Estimation (1996).

Disappointment is unmistakably the major purpose behind client rot, however, what causes client disappointment? Hill has made 5 distinct holes to clarify client rot:

- **Hole 1: The limited time hole;** Customers may misconstrue organization's advertising correspondences. They may get annoyed and free regard and trust for the organization.
- **Hole 2: The understanding hole;** Understanding hole creates, when administrators of the association don't have an exact comprehension of client's needs and needs. In the event that they don't have a clue about what's vital for the clients, it is to a great degree hard to fulfill client's needs.
- **Hole 3: The procedural hole;** Assuming that the organization has a full comprehension of what makes a difference most to the clients, it will at present neglect to convey consumer satisfaction in the event that it has not made an interpretation of clients' desires into fitting working methodology and frameworks.
- **Hole 4: The social hole;** Sometimes associations have clear strategies, which are very much co-ordinated to clients' needs and needs yet don't accomplish a reliably abnormal state of consumer satisfaction since staff are deficiently prepared or declined to pursue the methods to the letter at record-breaking.

- **Hole 5: The observation hole;** It is conceivable that holes 1-4 don't exist, however, clients are as yet disappointed. This is on the grounds that clients' impression of the execution of your association may contrast from the real world. Awful encounters from the past might have frame and demeanour against the organization and fixing this slip-up might take a long time. (Nigel Hill, Handbook of Consumer Satisfaction and Satisfaction estimation, 1996, 7).

Major reasons behind Measuring Consumer Satisfaction

In the event that you don't quantify consumer satisfaction, you can't distinguish unsatisfied clients that could beat or abandon your negative client audits. You additionally can't distinguish cheerful clients, you could initiate as evangelists or referrers. At last, you can't foresee or counteract client agitate to design proactively without measurements to examine.

1. It encourages you distinguish unsatisfied clients

You can't investigate troubled clients'input or make changes to your item or administration to satisfy them, on the off-chance that you don't realize clients are miserable in any case. Ensure you're sending CS studies, breaking down, and following up on negative client input, regardless of the fact that it is so difficult to hear, as your best need, so you can keep them from stirring, leaving negative client surveys, or cautioning against your image to companions and partners.

2. It causes you distinguish cheerful clients

If you don't gauge consumer satisfaction, you can't recognize your upbeat clients, who are discovering accomplishment with your item or administration. What's more, in the event that you aren't organizing client achievement, chances are, your organization isn't developing - at any rate, that is the thing that we found in a study of about 1,000 business.

3. It encourages you conjecture and work proactively

Without estimating CS and comparative client achievement measurements, it's hard for client achievement groups to design and advise needs. Without measurements to illuminate how solid (or unfortunate) your client base is, client achievement groups can't work with Sales to enhance client desires, with Product to join item criticism, or Marketing to enhance the conclusion to-end client encounter. Pioneers require information and patterns to conjecture group execution throughout the following month, quarter, and year to adjust methodology, if necessary. Thus, there you have it. Consumer satisfaction is essential since it encourages you take care of issues, avoid agitate, and recognize upbeat clients that can turn into your supporters and evangelists.

Measurements of Consumer Satisfaction

Hokanson (1995) is influenced by numerous elements, which incorporate cordial representatives, respectful workers, proficient representatives, and supportive representatives, exactness of charging, aggressive estimating, benefit quality, great esteem and fast service we focus on nine elements of consumer satisfaction, which are Location, Additional Services, Item Quality, Service Quality, Facilities, Reliability, Process, Value for Cash, Staff and Personnel Benefit.

Area

The area of any store is in every case imperative. Area can mean comfort and openness. Area can likewise allude to the quantity of stores in a specific land setting. As indicated by Martinéz-Ruiz *et al.* (2010:280), propose that once an area is close to the home, then exchange costs related with buy, for example, transport expenses and time spent are probably going to be diminished. As per Reilly (1931), who grew Reilly law of retail attraction, which suggests that individuals are attracted to bigger shopping in this way bigger urban areas will in general pull in a bigger number of clients to shop their than littler ones along these lines the requirement for stores to think about area, when setting up offices. This is additionally bolstered by Craig *et al.* (1984), who utilize the focal place hypothesis to clarify how individuals living far away are pulled in to bigger stores, which are midway situated in bigger shopping centers offering more gathering of merchandise and enterprises than those stores inside their very own region offering less products and ventures.

Extra Administrations

This measurement will comprise of four components enrolment card, parking garage, infant regions and conveyance of products. Martinéz-Ruiz *et al.* (2010) state that clients dependably search for comfort advantage in the cutting-edge condition. Extra administrations are basically essential in the retail business and assume a job in deciding consumer satisfaction through formation of accommodation. For instance, Grewal *et al.* (2002) agree that extra administrations like the accessibility of leaving can make comfort for clients with vehicles in this manner prompting a constructive outcome on consumer satisfaction. Other extra administrations, like the participation card/steadfastness card like-wise give access to limits and special products.

Item Quality

Item quality as per Garvin (1987) is portrayed utilizing eight properties, which are: Performance which alludes to an items essential working attributes. 9 Features: These are extra highlights which are otherwise called the extravagant accessories of the item. Conformance: Which is depicted as the degree to which an item will work appropriately over a predefined timeframe under expressed states of utilization. Reliability: The likelihood that will work legitimately over a predetermined timeframe under expressed states of utilization. Sturdiness: It is the measure of utilization a client escapes utilization of an item before it physically decays or until the point, when substitution is ideal. Workableness: The speed competency and cordiality of fix. Style: How an item bids to all the five faculties of a human. Client saw quality: Customers view of an items quality dependent on notoriety of the firm. For motivations behind this examination, we have utilized 3 components to gauge the element of item quality to be specific; item assortment, freshness of items and toughness. We pursue this line on the grounds that as indicated by Yuen and Chan (2010), not all components/qualities utilized recommended by Garvin (1987) are significant, while some are observed to be less quantifiable. For instance, the components of execution and highlights can't be utilized in this examination as a result of the trouble in isolating essential item qualities auxiliary attributes. Assortment, as per Dhar *et al.* (2001) encourages retailers to serve diverse tastes and inclinations of its customers. Dellaert *et al.* (1998) surrenders that assortment does enable retailers to draw in more shoppers as well as persuade them to buy more while at the store. In the event that a retailer gives more noteworthy assortment in item classes, it can enhance the comfort of procurement along these lines, expanding consumer satisfaction. Assortment item determination can like-wise help decrease the apparent costs like exertion and travel time.

Offices

Under this measurement, we utilize three components, like presentation, music and perfect and open air to quantify the impact of store offices on consumer satisfaction. As per Inman *et al.* (2009), a supermarket is depicted as a position of tangible upgrades, where customers find beautiful item shows combined with foods grown from the ground with flawlessly shown bundles of bites and ads covering the floor. They further contend that a few clients at that point utilize the in-store upgrades as signs to help them to remember what perishables they require and furthermore declare that specific shoppers enter shops without the aim of purchasing certain merchandise, however, 10 wind up purchasing a specific arrangement of products in light of the fact that the in-store boosts has activated unrecognized needs and wants prompting in-store basic leadership. This is upheld by Dioiri (2007) who states that the business outlets/general stôres offices are critical since they can impact or change the obtaining conduct of buyers; along these lines outrageous consideration must be seen, when settling on a choice on promoting on the grounds that 70 per cent of procurement choices are made amid shopping. Terblanche and Boshoff (2004) portray general store offices as the total of components that add to a wonderful shopping environment, for example, shop format/passageways that make it simple to move around, store neatness and very much molded item shows. In view of expanded rivalry in the market, retailers are being headed to enhance their path and show the board systems. This is being done to enhance organization offer of buyer buys and wallet. Burke (2005) declares that path the executives includes the viable situation of classes in the store walkways to enhance clients' shopping knowledge.

Unwavering Quality

Unwavering quality alludes to how much trust can be managed the general store staff and association, for instance through, parameters like precise charging. "Unwavering quality alludes to the guarantees given by the store. On the off-chance that the store can't keep or breaks the guarantees, it disappoints clients and results in negative informal. Interestingly, when the organization can stay faithful to its commitments, it expands client trust in the store and makes consumer satisfaction and lead to devotion", (Yuen and Chan, 2010:236) 2.2.6.

Process

Process will be estimated utilizing three components, for example, number of checkout counters/express checkout counters, opening hours and line holding up time at counters. It is critical to deal with these components of process in administration conveyance, as they can represent the moment of truth Customer fulfillment. As indicated by Katz *et al.* (1991), holding up time in a retail location is an ordeal that can prompt customer disappointment. Therefore, Hui *et al.* (1997) additionally affirm that the previous can contrarily influence the store support conduct of a client. Past research by Kumar *et al.* (1997) finds that hold up length desires impact fulfillment with the holding up experience subsequently the need to discover how clients desires for a hold up impact their support choice. In their help, Taylor (1994) like-wise finds that more prominent customer fulfillment is gotten with shorter holding up periods. Accordingly, there is requirement for general stores specifically to oversee long lines particularly amid pinnacle hours, when the traffic in the stores is substantial. 2.2.7.

Incentive for Cash

As indicated by Ciavolino and Dahlgaard (2007), value for cash is the apparent dimension of value with respect to the cost paid for an item or administration. Estimation of cash depends on aggressive valuing of items, limits granted to clients, and advancements. Cronin and Taylor (1992) guaranteed that consumer satisfaction isn't just influenced by client benefits yet additionally by cost and accommodation. Moreover, a few explores have been done the incentive for cash and the esteem appended to it by clients. These investigations like-wise call attention to the contrast among cost and quality and how they impact apparent esteem, consumer satisfaction and client conduct. Moreover, Zeithaml (1988b) says that, "client esteem for cash is the general evaluation of the utility of an item dependent on observations dependent on what is gotten and what is given". Quantitative investigations by Keaveny (1995) into exchanging conduct in administrations uncovered outcomes that the greater part of clients deserted in view of poor value recognition.

Staff

Representatives are essential for a company's showcasing system. In this paper, the creators make utilization of two components to be specific well-disposed supportive staff and knowledgeable & fast execution to clarify the staff measurement. As per Gwinner *et al.* (2005), Liao and Chuang (2004) state that the fruitful execution of a company's advertising idea is to a fundamental degree subject to the cutting edge representatives in view of their immediate client communication. While managing clients, workers are intended to be getting it done conduct. They are intended to act with a certain goal in mind and a few demonstrations are either positive or negative. Worker conduct is imperative in an administration organization since representatives go about as a connection for the association with its clients. Moreover, Gwinner *et al.* (1998) representatives speak to a basic factor in creating compelling working associations with clients. In their investigation, Lemmink and Mattson (1998) demonstrated that the level of individual warmth shown by administration representatives towards clients was connected fundamentally and decidedly to benefit quality recognitions and consumer satisfaction.

Faculty Benefit

A few specialists recommend that individuals like to associate outside their homes, for instance, Tauber (1972) declares that specific gatherings of individuals like o shop at specific stores on the grounds that the stores offer them chances to associate with individual customers and he additionally guarantees that customers like to shop at stores, where they discover cordial and affable faculty. Alludes to the nature of administration identified with the center items. Quality will be estimated utilizing factors like workforce benefit, benevolent staff, affable and educated staff. The speed of taking care of issues is additionally a critical variable. We additionally remember the capacity of staff to offer customize administration, for example, having the capacity to perceive visit clients and even welcome them by name.

Enhancing Customer Satisfaction in Retail

With consumer satisfaction falling in retail, it very well may entice roll out clearing improvements in your stores. In any case, steady minor enhancements can have as large an effect as considerable changes.

1. Legitimize the deal with social evidence

Late research demonstrates that customers purchase 100% of the time because of passionate reasons, not normal. After they have purchased, they need to legitimize their buy with rationale. They have to persuade themselves and the essential individuals in their lives that they've settled on a brilliant decision.

Regularly tributes get utilized in the business procedure. To take this to the following dimension and enhance consumer satisfaction, give a printed sheet of tributes or other social verification.

Customers would then be able to utilize this data to legitimize their choices.

2. Give customers a blessing, and afterward shock them with another

When you give customers a blessing, you initiate their correspondence reflex. This reflex was depicted in Robert Cialdini's book Influence: The Psychology of Persuasion. It is activated, when you give individuals something for nothing, and makes them feel committed to accomplish something consequently.

Giving blessings is a since a long time ago settled approach to support deals. Be that as it may, a test in eateries has demonstrated an approach to show signs of improvement results. For instance, give customers a blessing. Then restore a couple of minutes after the fact with a second one. Servers, who gave burger joints a mint and afterward returned unannounced and gave a second mint saw a 23% expansion in tips.

3. Take notes

There's a significant exercise in the self-improvement exemplary How to Win Friends and Influence People by Dale Carnegie. Everyone is the saint of his or her story. At the end of the day, everyone needs to feel critical.

For instance, your staff can make customers feel like they matter by esteeming what they need to state. Staff indicate they think clients are critical by focusing and taking notes, when customers are talking.

4. Train staff to deal with displeased clients

The way to taking care of bombshell clients originates from that equivalent Dale Carnegie rule about individuals. Individuals need to feel critical.

Here and there, workers dealing with the underlying grievance can make the client feel esteemed and take care of the issue. At different occasions, the main way your staff can do this is to heighten the issue to a senior representative.

When they do this, they should take care not to outrage the client further.

It's not in human instinct to be empathic towards somebody, when they assault us. That is the reason, you need to prepare staff to tackle the issue for the client notwithstanding, when under flame.

5. Have staff treat the clients as though they are the "manager"

At the point when representatives consider themselves to be working for you, they probably won't feel that they owe you an option that is other than the work, they are paid to perform. Show representatives another outlook, they are leader of their organization, an expert administrations firm. They work for themselves. You are basically their first, best customer.

In that situation, clients will profit by magnificent administration and your representative will push to expand their share. The representative's accomplishment in his organization gets attached to the capacity to enhance your consumer satisfaction.

6. Praise client birthday events

Does recalling client's birthday celebrations have any kind of effect? An overview demonstrates that this basic demonstration prompts 88% expansion in brand steadfastness.

7. Customization fulfills clients more than institutionalization

Clients search out tweaked encounters, as the world turns out to be progressively homogenous. On one hand, you need the experience clients have in your stores to be predictable. All things considered, you make your image through these encounters.

Nonetheless, you need to think of approaches to re-do the client encounter. Target did only that as of late, they sent a reference point innovation at their stores to draw in clients by and by while they peruse the paths.

8. Make a consumer satisfaction study

The reason you need to make a consumer satisfaction study is that it gives you a benchmark. It says, "This is the place we are presently."

Before you can get to where you need to be, you have to know your beginning stage.

Reviews enable you to keep tabs on your development, recognize issues and make a frictionless client encounter.

9. Encourage workers to be more administration arranged

Have you at any point gone into a store and solicited one from the staff to assist just with having them disclose to you, it isn't their activity?

Try not to commit a similar error in your stores. Train representatives that everybody is in client benefit. In the event that they can't tackle the client's concern, they should walk the client to the individual, who can.

10. Joy backhanded buyers

Much of the time, purchasing choices are affected in a roundabout way. At the point when customers are in the toy passageway, the individual who picks what to purchase is the kid.

Is it accurate to say that you are connecting with and pleasing this kind of backhanded purchaser?

11. Set up self-benefit checkouts

Barely any things are more baffling for customers than long queues. In numerous stores, a few checkouts are regularly not by any means open.

Why not transform your kept an eye on checkouts into self-serve checkouts? Have a staff part accessible to help individuals, when they buy and let clients submerge themselves in a 'hold up free' involvement.

12. Engage customers amid long holding up times

On the off-chance that your customers need to remain in line for quite a while, why not connect with them? Business is tied in with building connections, not shooting individuals with advertising messages. Why not contract a mystical performer, a comic, or even an artist to give some amusement amid this time? Simply keep it predictable with your image and engaging for your clients.

Keep in mind that feelings are in charge of acquiring choices, a great experience while holding up can lift client spirits and enhance their general understanding.

13. Set up a worker mentorship program

Numerous retailers neglect to take advantage of the learning and experience of their ranking staff. Every week, a lesser worker can get together with a senior colleague. In their time together, they can audit execution and even bond.

Representative bonds are significant for good. It's one of our most fundamental needs, we want to be a piece of something and feel acknowledged.

□□□

Chapter 5

Farm Fresh Retailing: *Case of Customer Satisfaction*

Introduction

Retail sector is the largest source of employment after agriculture, and has deep penetration into rural India. It is also believed that 21 million people are employed in the retail sector, which is 7 per cent of the total national workforce, whereas the global average is around 10-12 per cent. It is estimated that an additional eight million jobs will be generated through direct and indirect employment related to the retail sector (A. Sengupta, 2018).

Importance of the Study

It provides guideline for further research in area for organized retail. Research says about customer buying behaviour towards Big Bazaar. The research is also important to identify market size, growth and market Potential of Big Bazaar. The research shows future scenario of Big Bazaar in current perspective. The study shows opportunities and challenges for Big Bazaar with respect to internal & external environment. The study says about main competitors in the field of organized retail sectors and provides guidelines to further extension of Big Bazaar. The study provides help to know the customers satisfaction with Big Bazaar stores.

Objectives of the Study

- To study the satisfaction level of customers.
- To know the buying behaviour of the customers.
- To identify the main competitors of big bazaar with regard to services.

Scope of the Study

The scope of this research is to identify the buying behaviour and customer satisfaction of customers of Big Bazaar. This research is based on primary data and secondary data. The study is completely focused upon the objectives. The result, obtained from the objectives, will be helpful to the organization in knowing the different opinions of the customers about the company's products. The study will also be helpful for the company to take steps for solving the issues, related with problems in the service offered. In addition, this will be useful in finding out the valid steps to trigger the sales. Lastly, the study will also be a corner stone for improving ideas to bring up positive results in customer satisfaction.

Limitations of the Study

1. The study relates only to the consumers of area of Jammu city of J&K, therefore, the findings cannot be true representation of all consumers (total population).
2. The personal bias of respondents has also been a limitation.
3. Respondent's ignorance to certain question also posed as barrier towards certain responses.
4. Since the researcher has used structured schedule, the respondent's responses were immediate and the time taken for deciding was less, which led to responses with not much thought. Therefore, time constraint has also been a limitation in this study.
5. The study is only for the big bazaar confined to a particular location and a very small sample of respondents. Hence, the findings cannot be treated as representative of the entire retail industry.
6. Respondents may give biased answers for the required data. Some of the respondents did not like to respond.

In spite of the above said limitations, as a researcher I tried my best to extract more and more information with correctness.

Methodology

A methodology is usually a guideline system for solving a problem, with specific components, such as phases, tasks, methods, techniques and tools. The main goal of this study is to study the customer satisfaction towards the Big Bazaar in Jammu. The present investigation entitled **"A study on customer satisfaction with reference to retail outlet Big Bazaar in Jammu"** was carried out in 2017. The sampling structure and techniques adopted during the course of investigation have been described in this chapter.

Locale of Study

The research is conducted with the help of questionnaire, based on the information collected from the consumers of Big Bazaar in Jammu city.

Sample Size

The sample used in this study, consisted of the consumers in different areas of Jammu city. The data was collected through filling up the questionnaire from the consumers of Big Bazaar through market survey. The consumers, who made a visit to Big Bazaar, formed the sample unit and among them, 200 repondents were selected for the study purpose.

Data Collection

The data collection is the method to collect important information to keep on record for further use, to make important decisions about different issues and is of vital significance for others. The present study has adopted both the primary as well as the secondary data collection techniques to conduct the present study.

Primary Data

It involved the first hand information collected through questionnaires by direct interviews, helping the customers of Big Bazaar to fill up the questionnaires.

Variables Analysed

a) Demographics (gender, income, education, age and occupation)

b) Consumer satisfaction

Secondary Data

The data, which has already been collected, compiled and presented earlier by any agency, was used for purpose of investigation.

The data has been collected through various websites, journals related to fast food industry, research papers, journals related to brand and advertisement, newspapers, magazines, etc.

Methods of Analysing Data

After collecting the data, the results were analyzed by using descriptive statistics, percentage method and Garrett ranking technique. Descriptive statistics means mean, median, mode, standard deviation and is used to minimize large number of variables into meaningful and correlated form.

Percentage Analysis

Percentage refers to a special kind of ratio. It is used to make comparison between two or more series of data. They can be used to compare the relative items, the distribution of two or more series of data, since the percentage reduces everything to a common base and thereby allow meaningful comparisons to be made.

Percentage = (x/y) X (100)

Where, x = number of respondents respond

y = total number of respondents

Arithmetic mean

The arithmetic mean is calculated by adding a group of numbers and then dividing by the count of those numbers.

Mean = $\Sigma (x_1, x_2 \ldots x_n)/n$

Where, $x_1, x_2 \ldots x_n$ are the observation

n= number of respondents

Median

Median, which is the middle number of a group of numbers; that is, half the numbers have values that are greater than the median, and half the numbers have values that are less than the median.

$$\textbf{Median} = \frac{\sum(N+1)}{2} \text{ th item}$$

Standard Deviation

The standard deviation is a measure of how widely values are dispersed from the average value (the mean).

$$\textbf{Standard Deviation (ó)} = \frac{\sum (\text{x-Mean})^2}{(\text{N-1})}$$

Where, n is the sample size.

Mode

Mode is defined as the element, which has the highest frequency in a given data set. In grouped data, two kinds of mode: the Modal Class, or Class with the highest frequency and the mode itself can be find out, mode is calculated using the formula below:

$$\text{Mode} = L + \left[\frac{f_1 - f_0}{2f_1 - f_0 - f_2}\right] \times h$$

Where,

L is the lower class limit of the modal class

F_1 is the frequency of the modal class

F_0 is the frequency of the class before the modal class in the frequency table

F_2 is the frequency of the class after the modal class in the frequency table

h is the class interval of the modal class

Henry Garret Ranking Technique

In this technique, the percentage position of each rank obtained is converted into scores by referring to the table given by Henry Garret. Then for each factor, the scores of individual respondents are added together and divided by the total number of respondents for whom the scores are added.

Formula

Percentage position=100 (R_{ij}-0.5)/n

Where, R_{ij} is the rank

n= number of items

Results and Discussion

Demographics of the Respondents

Figure 5.1 and 5.2 represent the demographics of the respondents at Big Bazaar. It is clear from figure 5.1 and table 5.1 that out of 200 respondents, the monthly income of 43.5 per cent employees was in the range of 10,000-20,000, followed by income range of above 30,000 (28 per cent), followed by income range of below 10,000 (21.5 per cent), followed by income range of 20,000-30,000 (7 per cent). Regarding the qualification of the respondents, it is interpreted from figure 5.2 and table 5.2 that out of 200 respondents, 44.5 per cent were graduate, 39.5 per cent were post-graduate and 16 per cent were undergraduate.

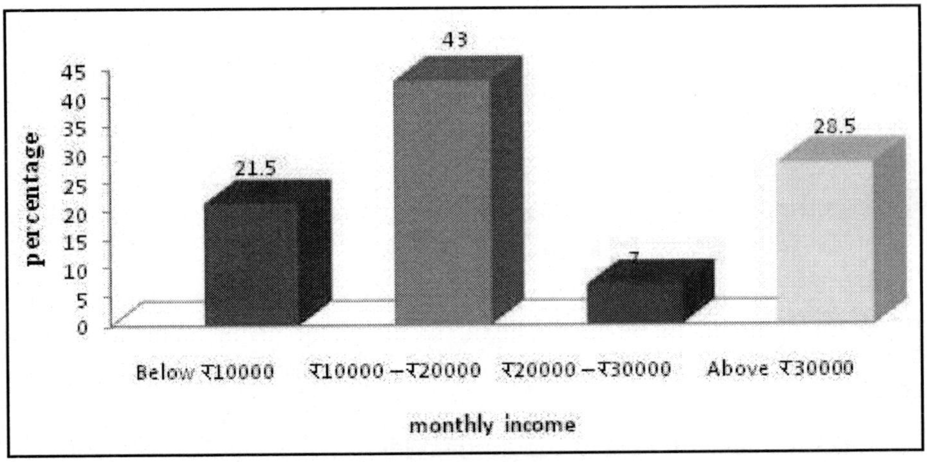

Fig. 5.1: Monthly income level

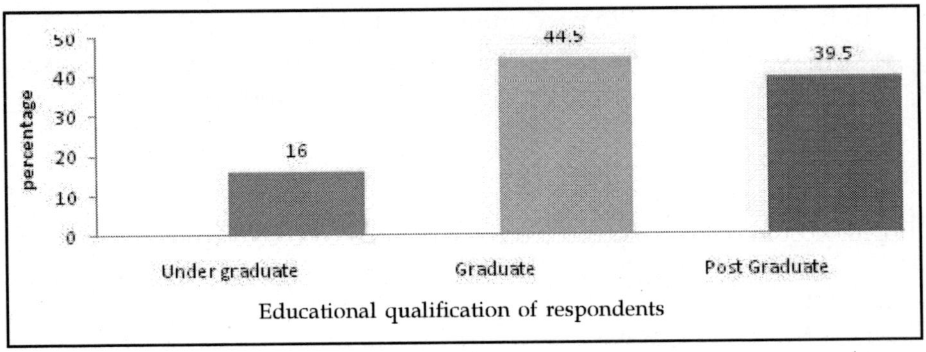

Fig. 5.2: Educational qualification

Table 5.1 : Monthly income level

Monthly income level	Frequency	Percentage
Below Rs. 10,000	43	21.50
Rs. 10,000 – Rs. 20,000	86	43.00
Rs. 20,000 – Rs. 30,000	14	7.00
Above Rs. 30,000	57	28.50

Table 5.2: Educational qualification

Educational Qualification	Frequency	Percentage
Under-Graduate	32	16.00
Graduate	89	44.50
Post-Graduate	79	39.50
Total	200	100.00

Frequency of visit to Big Bazaar

It is interpreted from the figure 5.3 and table 5.3 that frequency of visiting big bazaar was highest during special offers. Out of 200 respondents, 49 per cent visited the outlet during special offers, 35 per cent visited the outlet, whenever the need arose, and 14 per cent visited once a week and only 2 per cent visited twice a week.

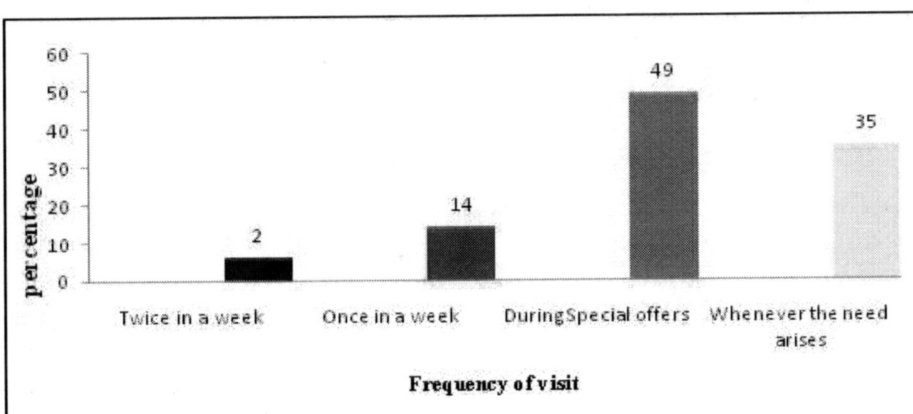

Fig. 5.3: Frequency of visit to big bazaar

Table 5.3: Frequency of visit to Big Bazaar

How frequently do you visit Big Bazaar	Frequency	Percentage
Twice in a week	4	2.00
Once in a week	28	14.00
During Special offers	98	49.00
Whenever the need arises	70	35.00
Total	200	100.00

Ranking of Factors in terms of Service

Figure 5.4 and table 5.4 shows the ranking of various factors in terms of service. Out of 200 respondents , 51 per cent said on time staff availability was good , 23 per cent said it was average , 12 per cent said it was fair,11 per cent said it was excellent and 3 per cent said it was poor. Regarding staff greeting and helping, 43 per cent said it was good, 24 per cent said it was average, 16 per cent said it was poor, 14 per cent said it was fairly good and 3 per cent said it was excellent. When the respondents were asked about the answering of questions by the staff, 39 per cent said it was good, 32 per cent said it was average, 13 per cent said excellent, 9 per cent said it was fair and 7 per cent said it was poor. Regarding the question of staff showing knowledge of products/services, 46 said it was average, 29 per cent said it was good, 10 per cent said it was poor, 8 per cent said it was fair and only 7per cent said it was excellent. Regarding to staff courtesy, 46 per cent said it was good, 22 per cent said it was excellent, 14 per cent said it was average, 10 per cent said it was fair and 8 per cent said it was poor. Related to overall customer service at Big Bazaar, 44.5 per cent rated it good, 22.5 per cent rated it excellent, 21 per cent rated it average, 8.5 per cent rated it fair and only 3.5 per cent rated it poor.

Preference of Consumers for Various Food Varieties

The figure 5.5 and table 5.5 interpret the preference of consumers for various food varieties. Out of 200 respondents, 19 per cent said they never preferred staple food in Big Bazaar, 9 per cent rarely preferred staples, 7 per cent sometimes preferred it, 43 per cent preferred it often and 22 per cent always preferred staples. Out of 200 respondents, 70.5 per cent often bought from farm fresh section and 29.5 always preferred it. 7.5 per cent out of 200 never preferred frozen foods,39.5 per cent rarely preferred frozen, 31.5 per cent sometimes preferred them, 19.5 per cent often preferred them and 2 per cent always preferred them.

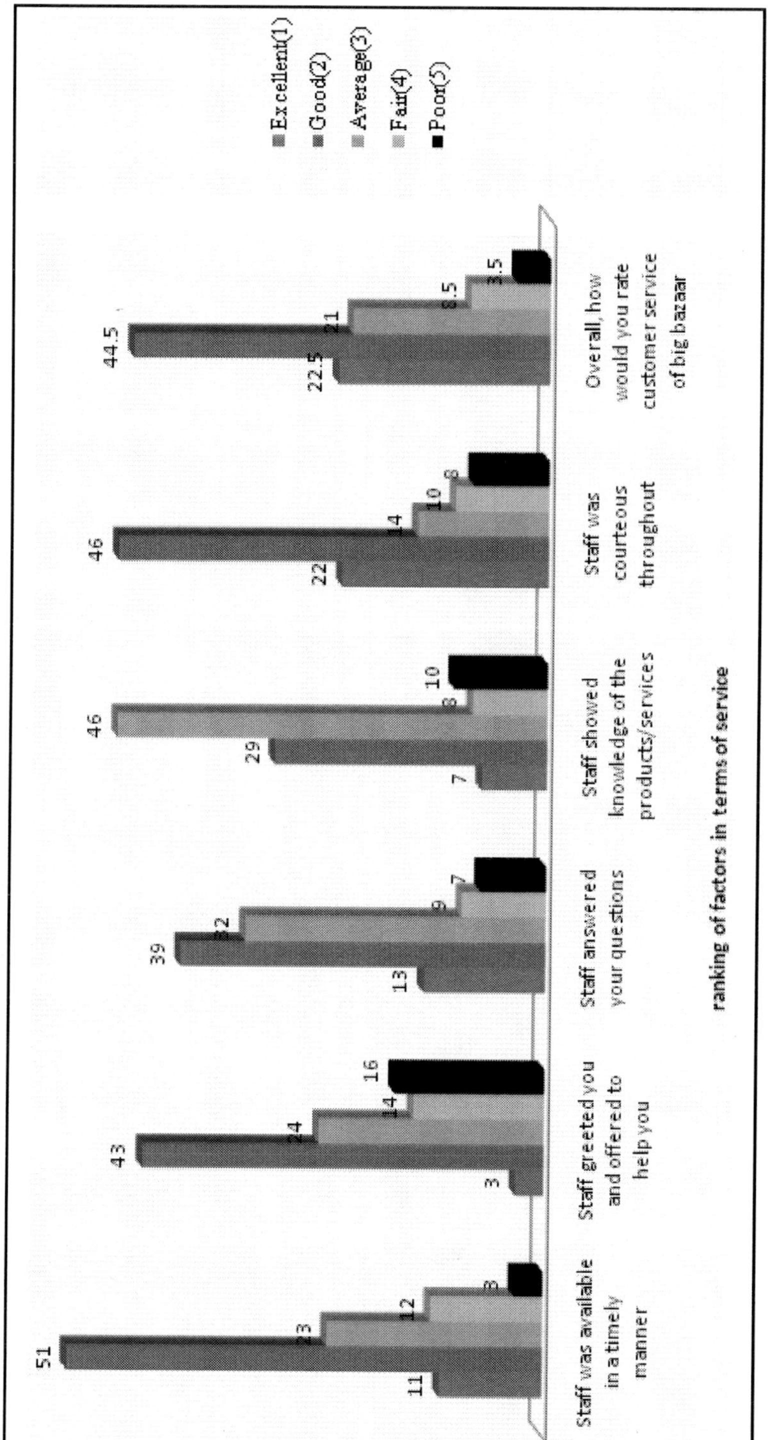

Fig. 5.4: Ranking of factors in terms of service

Table 5.4: Represents ranking of factors in terms of service

Rank the following factors in terms of service	Frequency						Percentage				
Factors	**Excellent (1)**	**Good (2)**	**Average (3)**	**Fair (4)**	**Poor (5)**	**Total**	**Excellent (1)**	**Good (2)**	**Average (3)**	**Fair (4)**	**Poor (5)**
Staff was available in a timely manner.	22	102	46	24	6	200	11.00	51.00	23.00	12.00	3.00
Staff greeted you and offered to help you.	6	86	48	28	32	200	3.00	43.00	24.00	14.00	16.00
Staff answered your questions.	26	78	64	18	14	200	13.00	39.00	32.00	9.00	7.00
Staff showed knowledge of the products/services.	14	58	92	16	20	200	7.00	29.00	46.00	8.00	10.00
Staff was courteous throughout.	44	92	28	20	16	200	22.00	46.00	14.00	10.00	8.00
Overall, how would you rate customer service of big bazaar.	45	89	42	17	7	200	22.50	44.50	21.00	8.50	3.50
Total	157	505	320	123	95						

Out of 200, 8.5 per cent never preferred food and beverage section, 24.5 per cent rarely preferred it, 46 per cent preferred it sometimes, 15.5 per cent preferred it often and 10.5 per cent preferred it always.

Recommendations by Consumers for Boosting Service

The figure 5.6 and table 5.6 represent the recommendation by consumers for boosting the service of big bazaar. For improving the

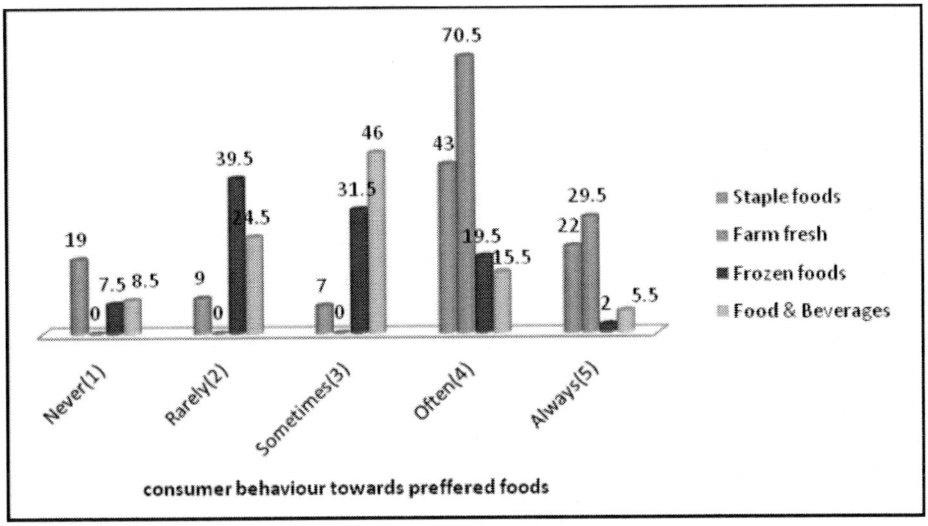

Fig 5.5: Consumer behaviour towards preferred foods from food varieties

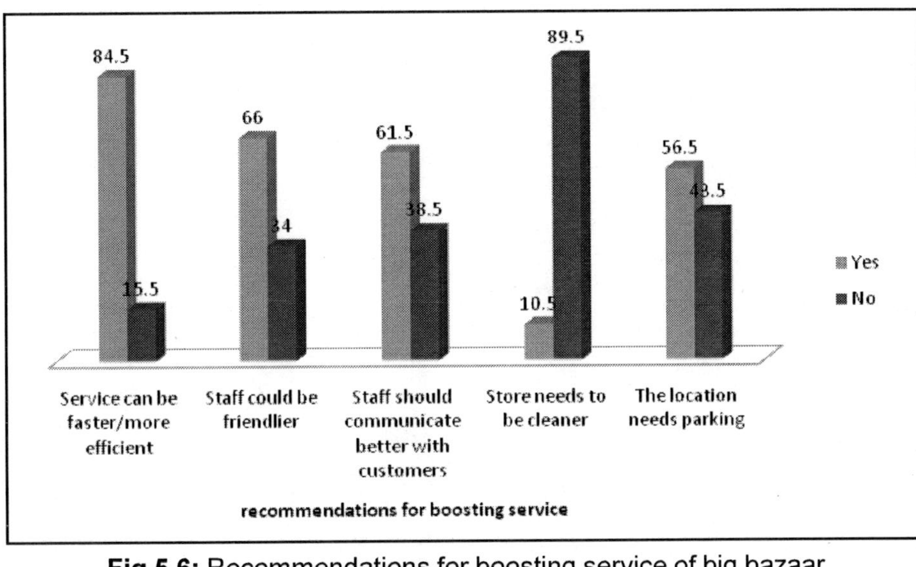

Fig 5.6: Recommendations for boosting service of big bazaar

Table 5.5: Represents consumer behaviour towards preferred foods from food varieties

Factors	Frequency						Percentage				
	Never (1)	Rarely (2)	Sometimes (3)	Often (4)	Always (5)	Total	Never (1)	Rarely (2)	Sometimes (3)	Often (4)	Always (5)
Staple foods	38	18	14	86	44	200	19	9	7	43	22
Farm fresh	0	0	0	141	59	200	0	0	0	70.5	29.5
Frozen foods	15	79	63	39	4	200	7.5	39.5	31.5	19.5	2
Food & Beverages	17	49	92	31	11	200	8.5	24.5	46	15.5	5.5

Table 5.6: Represents recommendations for boosting service of big bazaar

What would you like to recommend for boosting service of big bazaar			Total	Percentage	
Factors	Yes	No			
Service can be faster/more efficient	169	31	200	84.50	15.50
Staff could be friendlier	132	68	200	66.00	34.00
Staff should communicate better with customers	123	77	200	61.50	38.50
Store needs to be cleaner	21	179	200	10.50	89.50
The location needs parking	113	87	200	56.50	43.50

service or making it more efficient, 84.5 per cent said yes, it can be made faster, 15.5 per cent said no, there is no need. Regarding the question of staff could be friendlier, 66 per cent said yes, 34 per cent said no. Regarding the question of staff communication, 61.5 per cent said it can be better, 38. 5 per cent said no need. Regarding store cleanliness 89.5 per cent said it is clean, 10 .5 per cent said it should be cleaned. Regarding the question of whether store needs parking , 56.5 per cent said yes, whereas 43.5 per cent said no.

Satisfaction towards Food Section of Big Bazaar

The figure 5.7and table 5.7 represent the satisfaction level of consumers towards the food section of big bazaar. Regarding the price of food, 43.5 per cent said it was moderate, 36.5 per cent were satisfied, 20 per cent were dissatisfied. Regarding the quality of food, 48.5 per cent said the quality was moderate, 28.5 per cent said the quality was good, 18 per cent said they were dissatisfied, 4 per cent were highly satisfied and 1 per cent were highly dissatisfied. Among 200 respondents, 44.5 per cent said the taste of food was moderate, 25.5 per cent said the taste was good, 20.5 per cent said they were dissatisfied with the taste, 5.5 per cent were highly satisfied and 4 per cent were highly dissatisfied. Freshness of food was felt moderate by 47.5 per cent out of 200 respondents, Good by 31.5 per cent, 21 per cent were dissatisfied. Hospitality was found moderate by 42.5 per cent out of 200, Good by 39 per cent, 11.5 per cent were dissatisfied and 7 per cent were highly satisfied. The environment at Big Bazaar was found moderate by 48.5 per cent out of 200 respondents, Good by 38, 28 were dissatisfied, 29 found it excellent 8 were highly dissatisfied. By the attractiveness through advertisement, 58.5 per cent were moderately satisfied, 14.5 per cent were satisfied, 14.5 percent were highly satisfied, 13 per cent were highly dissatisfied, 1 per cent were dissatisfied and 4 were highly satisfied. With the variety of food available at the store, 64.5 per cent were moderately satisfied, 43 were satisfied, 11 per cent were dissatisfied and 6 were highly dissatisfied.

Preference of Retail Outlet

Figure 5.8 and table 5.8 show that when the respondents were asked about which retail outlet they prefer,43 per cent out of 200 said they preferred V-Mart,46 per cent said they preferred Big Bazaar and 11 per cent said they preferred Easy Day.

Table 5.7: Represents customer satisfaction towards food section ofBig bazaar

Rank the factors of Customer Satisfaction towards the food section of big bazaar		Frequency					Percentage				
Factors	Highly satisfied (5)	Satisfied (4)	Moderate (3)	Dissatisfied (2)	Highly dissatisfied (1)	Total	Highly satisfied (5)	Satisfied (4)	Moderate (3)	Dissatisfied (2)	Highly dissatisfied (1)
Price of food	0	73	87	40	0	200	0.00	36.50	43.50	20.00	0.00
Quality of food	8	57	97	36	2	200	4.00	28.50	48.50	18.00	1.00
Taste of food	11	51	89	41	8	200	5.50	25.50	44.50	20.50	4.00
Freshness of the food	0	63	95	42	0	200	0.00	31.50	47.50	21.00	0.00
Hospitality of the staff	14	78	85	23	0	200	7.00	39.00	42.50	11.50	0.00
Environment	29	38	97	28	8	200	14.40	29.00	48.50	14.00	4.00
Advertisement attractiveness		29	117	24	26	200	2.00	14.50	58.50	12.00	13.00
Variety of foods		43	129	22	6	200	0.00	21.50	64.50	11.00	3.00

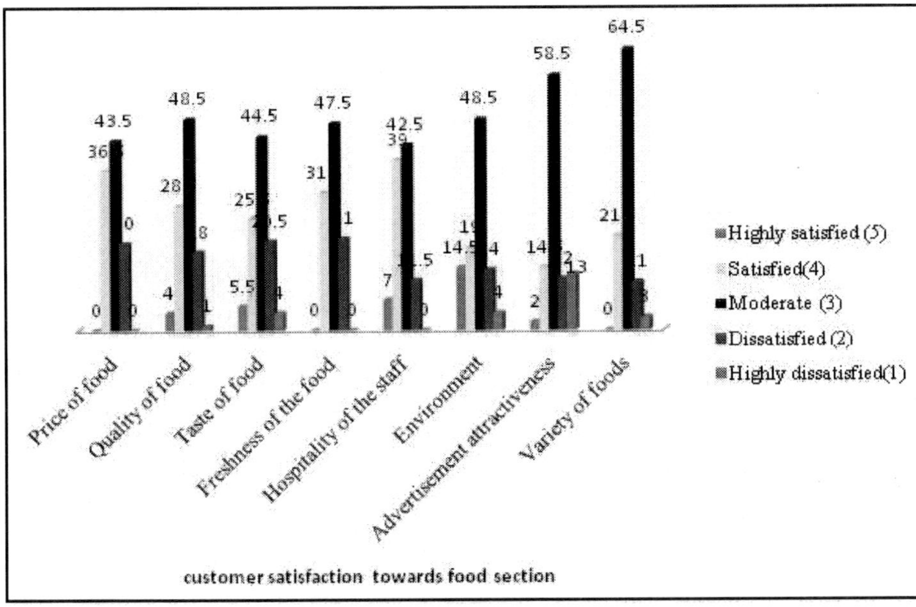

Fig 5.7: Customer satisfaction towards food section of Big Bazaar

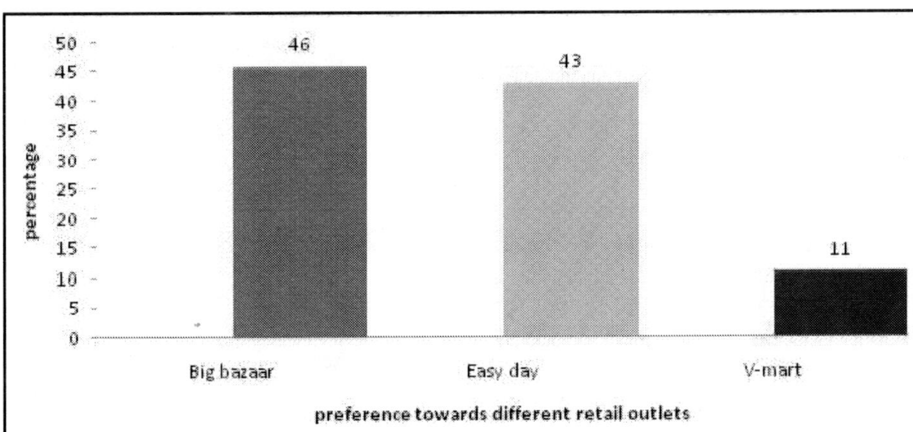

Fig.5.8: Preference towards different retail outlets

Table 5.8: Preference towards different retail outlets

Outlet	Frequency	Percentage
Big bazaar	92	46.00
Easy day	86	43.00
V-mart	22	11
Total	200	100

Frequency of Shopping

The figure 5.9 and table 5.9 represent the frequency of shopping by consumers to different retail outlets. Out of 200 respondents, 48.5 per cent said they visited retail outlets on special offers only, 26 per cent said they went for shopping to outlets monthly, 21.5 per cent said they visited outlets in festivals, 4 per cent said they visited the outlets weekly.

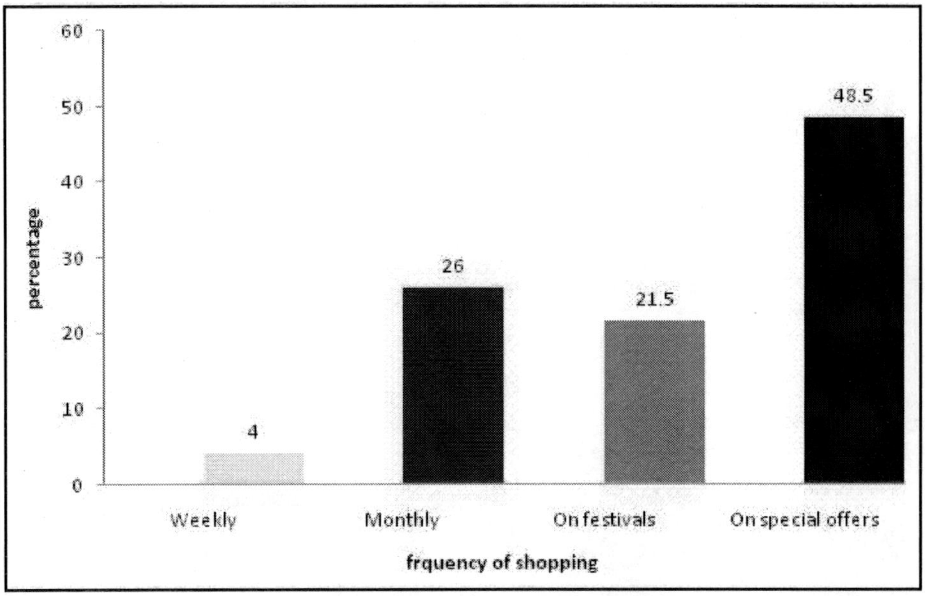

Fig. 5.9: Frequency of shopping by respondents

Table 5.9: Frequency of shopping by respondents

How often do you go shopping to retail outlets	Frequency	Percentage
Weekly	8	4.00
Monthly	52	26.00
On festivals	43	21.50
On special offers	97	48.50
Total	200	100.00

What Customers mainly buy from Big Bazaar

The figure 5.10 and table 5.10 answer the question, what do customers mainly buy from big bazaar? Out of 200 respondents, 49 per cent said they bought mainly vegetables, 34 per cent said they bought staple food, 11 per cent said they bought beverages and 6 said they bought frozen food.

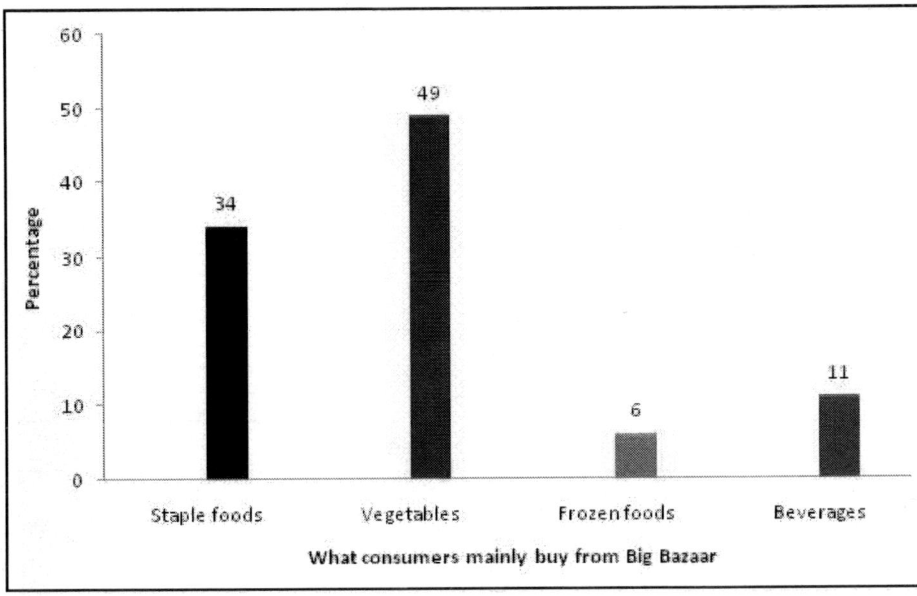

Fig. 5.10: What consumers mainly buy from Big bazaar

Table 5.10: What respondents mainly buy from big bazaar

Variables	Frequency	Percentage (%)
Staple foods	68	34.00
Vegetables	98	49.00
Frozen foods	12	6.00
Beverages	22	11.00
Total	200	100.00

Criteria for Choosing a Retail Outlet

The figure 5.11 and 5.11 shows the criteria for choosing a retail outlet. 60.5 per cent said they saw the price, 23.5 per cent said they saw product quality, 11per cent said they saw the location and only 5 per cent said they saw for product variety.

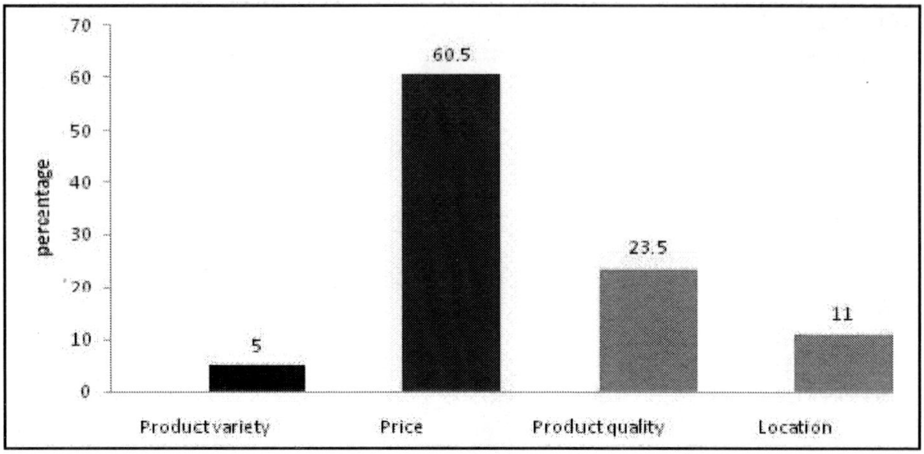

Fig 5.11: Criteria for choosing retail outlet

Table 5.11: Criteria for choosing retail outlet

Variables	Frequency	Percentage
Product variety	10	5.00
Price	121	60.50
Product quality	47	23.50
Location	22	11.00
Total	200	100.00

Response of Customers for Product Labelling

Figure 5.12 revealed on the question of considering product labelling while purchasing, 96 per cent respondents said yes and 4 per cent said no.

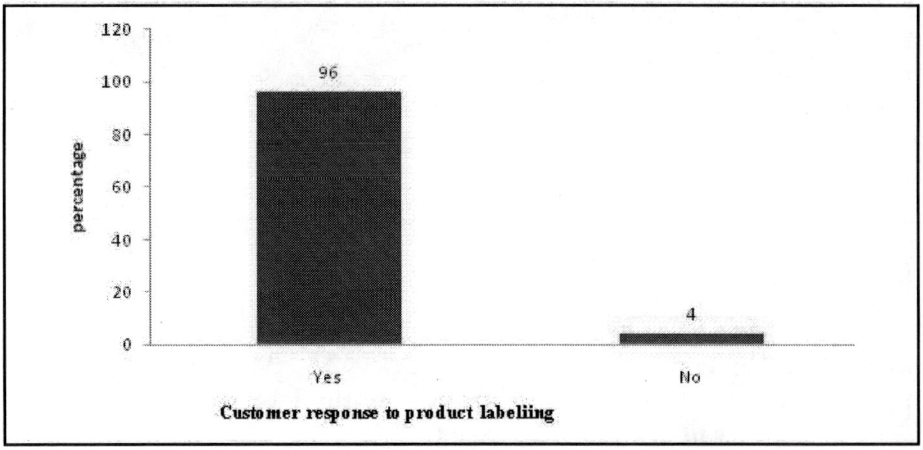

Fig 5.12: Customer response to product labelling

Table 5.12: Response to product labelling

Responses	Frequency	Percentage
Yes	192	96.00
No	08	4.00
Total	200	100.00

Ranking of Big Bazaar's Own Products

The figure 5.13 and table 5.13 show the ranking of big bazaar's own products. On the basis of price, 12 per cent ranked them poor, 59 per cent ranked it average, 29 per cent ranked it good.

Table 5.13: Rating of Big Bazaar's Own Products

Factors	Poor	Average	Good	Excellent
Price	12.00	59.00	29.00	0.00
Quality	5.50	56.00	34.50	4.00
Taste	16.00	44.50	38.00	1.50
Durability	17.00	48.50	29.50	5.00

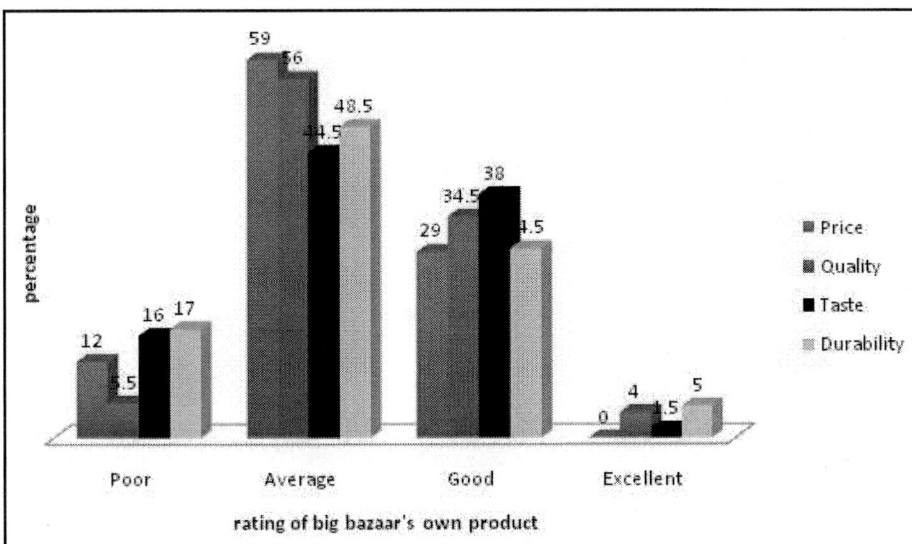

Fig 5.13: Rating of Big Bazaar's own products

On the basis of quality 5.5 per cent out of 200 respondents ranked big bazaar's own products poor, 56 per cent ranked them average, 34.5 per cent ranked them good, 4 per cent ranked them excellent. On the basis of taste, 16 per cent rated them poor, 44.5 per cent rated them average, 38 per cent ranked them good and 2.5 per cent ranked them excellent, regarding the durability of Big Bazaar's own products, 17 per cent ranked them poor, 48.5 per cent ranked them average, 29.50 ranked them good and 10 ranked them excellent.

Comparison of Big Bazaar with other Retail Outlets

The figure 5.14 and table 5.14 represents the comparison of big bazaar with other retail outlets on the basis of product variety, 21 per

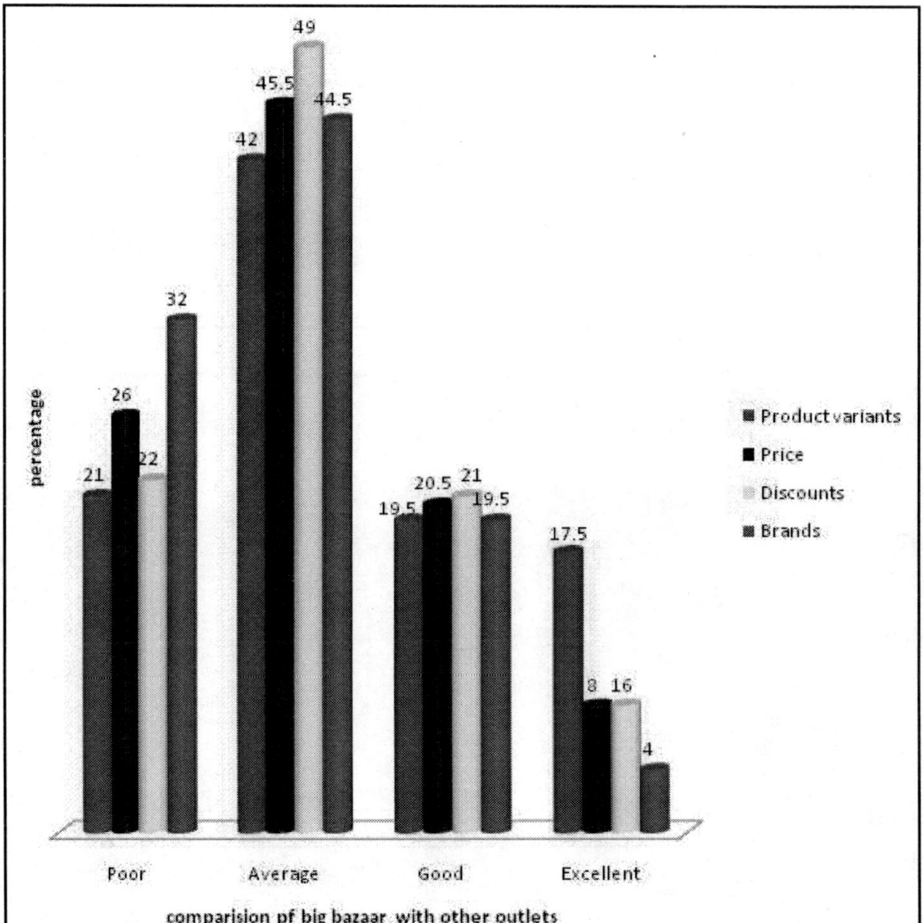

Fig 5.14: Comparison of Big Bazaar with other retail outlets

cent rated it poor, 42 per cent rated it average,19.5 per cent rated it good, and 17.5 per cent rated it excellent. On the basis of price, 26 per cent rated it poor, 45.5 per cent rated it average, 20..5 per cent rated it good, 16 rated it excellent. On the basis of discounts given 22 per cent rated it poor, 49 per cent rated it. 11 per cent rated it good and 18 per cent rated it excellent. On the basis of brands available, 32 per cent rated it poor, 44.5 per cent rated it average, 19.5 per cent rated it good and 4 per cent rated it excellent.

Table 5.14: Comparison of Big Bazaar with other retail outlets

Factors	Poor	Average	Good	Excellent
Product variants	21.00	42.00	19.50	17.50
Price	26.00	45.50	20.50	8.00
Discounts	22.00	49.00	21.00	8.00
Brands	32.00	44.50	19.50	4.00

Ranking of Primary Factors in terms of Service

Table 5.15 represents the ranking of various factors in terms of service. Using garret ranking, overall customer service of big bazaar was ranked highest, followed by staff courteously, followed by timely staff availability, followed by answering questions by staff followed by staff showing the knowledge of products, followed by staff greeting and offering help.

Ranking of other Service Factors

Table 5.16 represents the ranking of service factors in big bazaar. The store layout makes customers to find easily what they need was ranked highest, followed by the service factor the store layout makes customers easy to move around, followed by staff communication in understandable language, followed by the service factor that staff delivers appropriate service as promised, followed by service factor that complaints are constructively handled, followed by big bazaar accepts majority of credit cards, followed by service factor that big bazaar provides plenty of convenient parking.

Table 5.15: Rankingof primary factors in terms of service

Factors	Excellent (1*77)	Good (2*63)	Average (3*54)	Fair (4*46)	Poor (5*37)	Total	Average score	Rank
Staff was available in a timely manner.	1694	6426	2484	1104	222	11930	59.65	III
Staff greeted you and offered to help you.	462	5418	2592	1288	1184	10944	54.72	VI
Staff answered your questions.	2002	4914	3456	828	518	11718	58.59	IV
Staff showed knowledge of the products/services.	1078	3654	4968	736	740	11176	55.88	V
Staff was courteous throughout.	3388	5796	1512	920	592	12208	61.04	II
Overall, how would you rate customer service of big bazaar.	3465	5607	2268	782	259	12381	61.90	I

Table 5.16: Ranking of other service factors

Factors	Strongly disagree (5*47)	Disagree (4*53)	Neither agree nor disagree (3*60)	Agree (2*68)	Strongly agree (1*80)	Total	Average score	Rank
Your complaints are constructively handled.	282	1272	5340	4828	800	12522	62.61	V
Staff delivers the appropriate service, as promised.	47	3551	120	6664	2560	12942	64.71	IV
Staff communicates in a language that you understand.	564	1537	1680	7276	1920	12977	64.88	III
The behaviour of staff makes you feel that you can trust them and have confidence in them.	0	3657	2820	5712	0	12189	60.94	VII
The store layout at this store makes it easy for customers to find what they need.	0	106	720	9316	3920	14062	70.31	I
The store layout at this store makes it easy for customers to move around the store.	0	0	420	11492	1920	13832	69.16	II
Big bazaar provides plenty of convenient parking for customers.	141	5883	900	3604	1440	11968	59.84	VIII
Big bazaar accepts most major credit cards.	893	2332	2280	5848	1040	12393	61.96	VI

□□□

Chapter 6

Summary and Conclusion

Summary

The information was collected from the sample area Jammu city. The data collected was subject to analysis for examining the objectives of investigation. The finding of the present investigation has been briefly summarized in this chapter.

To know the satisfaction level of customers.

Big bazaar is an Indian retail store that operates as a chain of hypermarkets, discount department stores and grocery stores. Among all the factors, the respondents have picked quality and price the most important factors affecting customer satisfaction. Among all the factors, the respondents have picked price of food as the most important factor affecting customer satisfaction. Customers were mainly satisfied with the price of the commodities in the store. It is revealed from the study that the satisfaction level of customers mainly depends upon price of the products and their quality.

To know the buying behaviour of the customers.

Consumer buying behaviour includes the actions, thought process and perceived outcome during the course of making a decision, which results in a purchase. It is revealed from the study that consumers preferred to buy from the farm fresh section (70 per cent). It was also revealed from the study that maximum of the customers prefer to shop on special offers (48.5 per cent).

To identify the main competitors of Big Bazaar with regard to service

Big Bazaar is not the only retailer in the market. It faces huge competition from other players of the market, such as Wal-Mart, Vishal Mega Mart and other local retailers. It was revealed from the study that only 19.2 per cent rated Big Bazaar as good as compared to its competitors on the basis of product variety, thus faces huge competition in product variety as other outlets, such as Vishal Mega Mart, etc. have more diversity in products. But it was also revealed found that Big Bazaar has lesser competition due to the reason that it provides time-to-time offers, such as on festivals, holidays, etc.

Conclusion

The concept of everything under one umbrella and competitive prices brings more customers to the store. Big bazaar provides quality products at low price and thus, it justifies the statement"iss se sasta aur achha kahin nahi!"

The present study was carried out with the main aim to study the customer satisfaction towards the outlet. The main aim of the study was to know the attributes and factors influencing consumer satisfaction towards big bazaar in Jammu. It is revealed from the study.

- That maximum of the respondents were in the income range of 10,000-20,000 and mostly the respondents were graduates.
- Frequency of visiting the outlet was highest during special offers.
- Regarding the various service factors, it was revealed from the study that staff availability was good, staff answered well to the questions of the customers. Regarding the staff courtesy, the response of maximum respondents was good.
- It is revealed from the study that complaint handling was good as response of most of the respondents was positive with this regard. Regarding the communication, maximum of the respondents were of the view that staff communicates in a language understood by customers.
- It is revealed from the study that maximum of the customers come to Big Bazaar for buying fruits and vegetables from farm fresh section .Frozen food was preferred the least.

- Regarding the food section of big bazaar, maximum customers told that they were moderately satisfied with price, freshness and taste of the food.
- It is revealed from the study that 96 per cent respondents see labelling before buying a product.
- About 60.5 per cent respondents chose the outlet on the basis of price of products available.

Recommendation

- Company should adopt the strategy to make good relations with customers by satisfying their requirements, understanding their preferences, etc.
- Company should focus on product quality. Customers were of the view that product quality of Big Bazaar's own products is not good. Thus, it can work on the product quality to increase sales and customer satisfaction.
- Company should work on customer segmentation and targeting. Big bazaar should segment its customers as per their needs so that they can be targeted with products, which they need. This will increase sales.
- The company can carry out more promotional activities. The study revealed that maximum promotion was through hoardings. Other promotional tools can be used.
- It should make different cash counters for different customers. Cash counter and credit card payment counter should be placed differently in order to reduce the rush and save the customers' time.

Limitations of the Study

- The study only relates to the customers of Jammu city of J&K, therefore, the findings cannot be the true representation of all the customers.
- The personal bias of respondents has also been a limitation.

- It should make different cash counters for different customers. Cash counter and credit card payment counter should be placed differently in order to reduce the rush and save the customers' time.
- Respondent's ignorance to certain questions also posed a barrier towards certain responses.
- Respondents tried to escape some statements by simply answering "neither agree nor disagree" to most of the statements.

Bibliography

Aghion, P., N. Bloom, R. Blundell, R. Griffith, and P. Howitt, 2005. "Competition and Innovation: An Inverted-U Relationship." *Quarterly Journal of Economics*, **120**(2): 701-728.

Aggarwal V., 2007. The Era of Retail Revolution: Contribution to Economy, Research *in Management and Technology*, Deep and Deep Publication: 429-442.

Anton, J., 1996. Customer Relationship Management: Making Hard Decisions with Soft numbers. Prentice-Hall, Englewood Cliffs, NJ: 73.

Anderson, E.W., Fornell, C., Lehmann, D.R., 1994. "Customer Satisfaction, Market Share and Profitability: Findings from Sweden, " *Journal of Marketing*, **58**(3): 53-66.

Arutselvi, M., 2012. "A Study on Consumers' Preference Towards Various Types of Britannia Biscuits in Kanchipuram Towns", *International Journal of Research in IT, Management and Engineering*, **2**(5): 1-13.

Assael and Henry, 1994. "Consumer behaviour and marketing action", 5th ed, South-Western Collage Publishing, Cincinnati, Ohio, USA.

Berman B and Evans J.R., 2008. Retail Management, Pearson education, Inc, 10th ed.

Berry, L.L. and Parasuraman, A., 1991. Marketing Services: Competing through Quality. The Free Press, New York, NY: 31.

Blackwell, R.D., Miniard, P.W. and Engel, J.F., 2001. "Consumer behaviour", Orlando: Harcourt College Publishers.

Blackwell, R.D., Miniard, P.W. & Engel, J.F., 2006. Consumer behaviour, l0th ed. Mason, OH: Thomson South-Western.

David. I.L. & Bitta A.J., Engel J.F., Blackwell R.D. and Miniard P.W., 1986. Consumer Behaviour, 5th ed.

Day G.S., 1998. What does it mean to be market-driven? Business Strategy Review, **9**(1):1-14.

East R., Wright M. and Vanhuele M., 2008. Consumer Behaviour: Applications in Marketing. London: Sage.

Engel, J.F., Blackwell, R.D., and Miniard, P.W., 1993. Consumer Behaviour, 5th edition. Dryden.

Engel J.F., Kollat D.T. & Blackwell R.D., 1968. Consumer behaviour, Holt, Rinehart, Winston, New York.

Fornell, C., Johnson, M.D., Anderson, E.W., Cha. J. & Bryant, B.E., 1996. The American Customer Satisfaction Index: Nature, Purpose, and Findings. *Journal of Marketing*, **60**(4): 7-18.

Hawkins, D.I. Best, R.J. and Coney, K.A., 2004. Consumer Behaviour Building Marketing Strategy, 9th ed, McGraw-Hill, New York, NY.

Harold, 1929. "Stability in Competition." *The Economic Journal*. **39**(153): 41-57.

Haubl, G., 2000. The Effects of Interactive Decision Aids. Consumer Decision Making in Online Shopping Environments: **5**(2): 223-227.

Ian C. MacMillan, Alexander B. van Putten, and Rita Gunther McGrath, 2003. "Global Games Manship. "*Harvard Business Review*, **81**(2): 62-71.

Jegan, A. & Sudalaiyandi S., 2013 "A Study on Consumer Buying Behaviour Towards Various Types of Sunfeast (ITC) Biscuits In Kovilpatti", *International Journal of Current research*, **5**(4): 957-960.

Karni A., G. Meyer, C. Rey Hipolito, P. Jezzard, M.M. Adams, R. Turner and L.G. Unger eider, 1998. The acquisition of skilled motor performance: fast and slow experience-driven changes in primary motor cortex, Proc. Natl. *Acad. Sci. U. S. A.* 95: 861–868.

Kapferer, J., 2008. "The New Strategic Brand Management", 4th edition, Kogan Page.

Kotler, P., 1999. Marketing Management 10th edition, Millennium Edition, Prentice Hall of India Private Limited New Delhi, 1-876.

Kotler, P., Wong, V., Saunders, J. & Armstrong, G., 2005. Principles of Marketing. Essex: Pearson Education Limited.

Kotler, P. & Keller, K., 2006. "Marketing Management", Twelfth edition, Prentice-Hall: 1-718.

Kotler P.A., Gary, Wong, Veronica and Saunders, John, 2008. Principles of Marketing, Fifth European edition, Pearson Education Limited. 265-271.

Kotler Philip, 2009. Marketing Management. A South Indian Perspective, 13th edition, Pearson Prentice Hall Publication.

Krugman, H., 1965, "The impact of television in advertising: learning without involvement", *Public Opinion Quarterly*, **30**(1): 349-56.

Kuester, Sabine, 2012. Marketing 301: Strategic Marketing & Marketing in Specific Industry Contexts, University of Mannheim: 110.

Kurniawan, B., 2010. Factors affecting customer satisfaction in purchase Decision on ticket online: a case study in air asia, **2**(1): 334-349.

Lim, H., 2010. Understanding American customer perceptions on Japanese food and services in the U.S. *UNLV Thesis/Dissertations/Professional Papers/ Capstones.*

Luo, Y., 1999. Environment-strategy-performance Relations in Small Business in China. *Journal of Small Business Management*, **37**: 37-52.

Mathew Joseph, Nirupama Sundararajan, Manisha Gupta, Sanghamitra Sahu, 2008. Impact of Organized Retailing on the Unorganized Sector. *Indian Council for Research on International Economic Relations.* **2**(6): 112-125.

McKinsey & Company, Inc. 2015. Retailing in India-The Emerging Revolution, CII.

Mishra. S., 2008. New Retail Models in India: Strategic Perspective Analysis, *Journal of Marketing & Communication*, **4**(2): 39-47.

Mohsan, F., Nawaz, M.M., Khan, M.S., Shallkat, Z. and Aslam, N., 2011. Impact of customer satisfaction on customer loyalty and intentions to switch: Evidence from banking sector of Pakistan. *International Journal of business and social science*, **2**(1): 16.

Oliver, R.L., 1981. Measurement and evaluation of satisfaction processes in retail settings. *Journal of Retailing*, **75**(2): 25-48.

Oliver, R.L., 1993. A conceptual model of service quality and service satisfaction: Compatible goals and different concepts, In Swart, T.A., Bowen, D.E., and Brown, S.W. (eds.).

Porter, M.E., 2004. "Competitive strategy: Techniques for analyzing industries and competitors", London: Free Press Essex: Pearson Education Limited.

Rajasekaran, B. and Saravanan, P.A., 2014. "Consumer Satisfaction on Fast Moving Consumer Goods", *GJRA - Global Journal for Research Analysis*, **2**(8): 38-41.

Ranaweera, C. and Prabhu, J., 2003. The influence of satisfaction, trust and switching barriers on customer retention in a continuous purchasing setting. *International journal of service industry management*, **14**(1): 37-95.

Rees, A.M., 1992. Factors influencing consumer choice. *Journal Soc. of Dairy Tech*, **45**(4): 112-116.

Saravanan, S., 2010. "A study on consumer behaviour of women with special reference todurable goods in Coimbatore city, Tamil Nadu", *Ind. Journal of Marketing*, **40**(1): 115-119.

Schiffman. L.G. & Kanuk, L.L., 2004. Consumer Behaviour. Prentice Hall of India, 3rd ed.

Sengupta, A., 2008. Emergence of modern Indian retail: an historical perspective, *international Journal of Retail & Distribution Management*, **36**(9): 689-700.

Shapiro, Carl. 2012. "Competition and Innovation: Did Arrow Hit the Bull's Eye" in the rate & direction of Economic activity revisited, Josh Lerner and Scott Stern, eds. Chicago: University of Chicago Press.

Singh, K., 2014. Retail sector in India: Present scenario, emerging opportunities and challenges. *IOSR Journal of Business and Management*. **4**(7): 121-129.

Solomon, Bamossy, Askegaard, K. and Hogg, 2006. "Consumer Behaviour, A European Perspective", 3th edition.

Srinivasan, N. and Elangovan, D. 2000. "Consumer perception towards processed fruits and vegetable products", *Indian Journal of Marketing*, **30**(11-12): 22-25.

Srivastava, R.K., 2008. Changing retail scene in India, International Journal of Retail andDistribution Management, **36**(9): 714-721.

Thorsten, H.T. & Alexander, K., 1997. The impact of customer satisfaction and relationship quality on customer retention: A critical reassessment and model development. *Psychology & Marketing*, **14**(8): 737-764.

Tse, D.K. & Wilton, P.C., 1988. "Models of Consumer Satisfaction: An Extension, " *Journal of Marketing Research*: 204-212.

Vavra, T.G., 1997. Improving your Measurement of Customer Satisfaction: A guide to creating, conducting, analysing, and reporting Customer Satisfaction Measurement Programs. American Society for Quality, **4** (5): 52-67.

Westbrook and Reilly, 1983. "The Vicious Circle of Consumer Complaints", *Journal of Marketing*, 48: 256.

Woodruff, R.B. & Gardial, S., 1996. Know Your Customer: New Approaches to Understanding Customer Value and Satisfaction. Wiley: Blackwell.

Zairi, M., 2000. Managing customer dissatisfaction through effective complaint management systems. *The TQM Magazine*, **12**(5): 331-335.

Zeithaml, P.A., 1985. A conceptual model of service quality and its implication for future research. *Journal of Marketing Research*, **49**(1): 41-50.

Zeithaml, V.A., Berry, L.L. and Parasuraman, A., 1996. The behavioural consequences of service quality. *Journal of Marketing*, **60**(8): 31-46.

Zeithaml, V.A. & Bitner, M.J., 2003. Service Marketing: Integrating Customer Focus across the Firm, New York: McGraw-Hill Higher Education.

□□□